Embracing the System

The Systemisation of Business and Life

by

Andrew Macfarlane

1stBooks – rev. 11/01/01

To Fiona, Hannah and Joel
who give me meaning beyond systems

Contents

Introduction

In the last decade of the twentieth century, a computer program defeated the world chess champion and groups of lawyers could not distinguish between computer generated summaries of law journal articles and those produced by lawyers. International trade grew to the extent that foreign exchange markets traded up to a trillion (million million) dollars per day. Yet the total amount of goods traded was about one twentieth of this and the dominant flow of money was pension funds and investments. Are the first two achievements history without precedent? The global financial trade requires a change in perspective which is difficult for most people. The techniques behind the computer accomplishments augur the emergence of a privileged class of super-specialists issuing arcane technologies which disrupt our perceptions of ourselves. We have entered an era where the complexity and size of social structure requires techniques in which the domain of intelligent judgement is invaded and replaced by calculations of extreme sophistication. On its own, this is an unprecedented challenge to humanity. We entered this stream of history in the closing decades of the nineteenth century, before the science which produced this challenge was invented. Its invention is a consequence of the systemisation.

For the prosperous of the Earth, the shape of our society is dominated by systems. We have cars, extraordinarily smooth roads, water free of harmful micro-organisms, flush toilets, an over-abundance of food and entertainment and, for many of us, too much to do. The history of how this came into being is the history of the industrial revolution, the history of commerce, the history of education and the history of science. This book concentrates on something which threads throughout all these separate histories. It is the concept of a system.

Today the word "system" elicits the image of information systems. This response has more to do with the success of powerful commercial concerns rather than what systems are. This book focuses on the idea of systems as things which transform inputs into outputs. If this is reliable we have a system. Thus clocks, locomotives, aircraft are systems. But, above all, the systems which have transformed our lives are the great manufacturing systems which grew up in the late nineteenth century and throughout the twentieth century.

It is argued that history has no sense of directions. But recent history has at least two: the accumulation of science and the movement towards growth and integration of systems. The two are interwoven but not identical. It is systems of all types which deliver to us the results of science. And it is science that provides the ability to produce the abundance of goods which is the shape of our society. It is this abundance which is the background to all the large social movements of recent history. Military histories detail battles but not the factories and social arrangements which produce the abundance of weapons. Histories of entertainment seldom detail the development of cameras, projectors, tape-recorders or the mass production of plastic records, lasers, compact discs, microprocessors and videos which have been part and parcel of the mass market for cinema, pop-music and videos.

This book is little more than an introduction to one of the biggest transformations of humanity - the application of systems to re-order life from the agricultural society of Thomas Jefferson or Jane Austin to a society of "heroic consumption" characterised by say Donald Trump or Bill Gates. Computers appear at the end of this story and were never, and have never been, crucial to it. Henry Ford is more characteristic of the twentieth century than a dozen computer pioneers. The growth of large-scale manufacturing and world trade has meant the growth of management and administration. Management and administration are about co-ordination, about holding the line against caprice, mischance and waste and have to deal with large numbers of variable entities all with a potential for disorder. They therefore require a tool to manage this variability. Hence their love of information technology. It is administration, whether corporate or government, whose job it is to create business and social order which has bonded the idea of system to that of information system. Administrators are systems makers; they must be, otherwise their judgements and decisions appear arbitrary. Dealing with a great variety of situations in a socially acceptable way requires large numbers of rules to cover these situations. Applying these rules must be justified by various facts. And information systems are just the devices to provide such easily presented facts. When this is successful, as it usually is, success breeds success until social processes are precisely described and proscribed. But precision brings complexity, a dark character which stalks the pages of this book. The greater the complexity, the more difficult it is to make universally acceptable decisions. Living with complexity, a result of the integration of commercial, financial and social systems, is now a constant characteristic of our society.

What this means, how it affects our social concepts, is explored in this book. But the book is not about computers or information technology as such. Inevitably, because they are the assumed tool of integration on a world scale, they must be discussed. A discussion of information technology and its effects in business and life can degenerate into a meal of opinion, statistics, sociology plus obligatory sauces and toppings of anxieties about instrumental thinking or unequal access and political dominance. While no book can be free of opinion, the direction taken here is to look at information technology, and information generally, in terms of systems or sequences of systems. Underlying this book is a vision of the way systems do work, can work and might work in the future. Every author has a slant with which to approach a topic and should "confess" earlier rather than later. The deeper thesis of the book is that much of what we see today relies on techniques that are increasingly abstract. That while engineering, which is based on physical principles, has produced the prosperity which characterises our society, management, including both marketing and financial, will use techniques that are based on statistical and logical principles. This is already the case as administrative databases rest on the foundation of modern abstract logic. Financial techniques use sophisticated statistical mathematics which will also invade marketing and business operations as they come to resemble commodity trading. The management of capability - the ability of an organisation to change - which determines the value of a company, will produce new problems and concepts that will force the development of new techniques. Furthermore, administration, control and information processing extend their compass by deploying these methods to define options and actions until a series of overlapping systems defines rationality. Add to this the increasing size and complexity of these overlapping systems and we are forced into another round of using abstract techniques to further automate higher levels of expertise to manage the creatures we have created.

The existence of mass production, not only of goods but of systems themselves, has changed the nature of employment. Manufacturing is highly systematised, very competitive and necessarily innovative. This produces rapid obsolescence in skills and has fed the decline of manufacturing as a source of careers. This, in turn, has produced a nervous search for durable skills. A search which is now considered to be in vain. As the education of India and China proceeds apace and the movement of people around the world becomes easier, skills will become a commodity market. This book looks at the types of skills which are subject to automation and why some skills are immune.

The production of complex systems requires systems themselves. In engineering, there are extensive checks to make sure all aspects of design are vetted and tested. These systems and checks have been developed so that some of the systems, which are innocently and quietly contributing to our well being, run safely. These potentially dangerous systems are part of the social fabric of modern society and depend on meticulous attention to detail for their good operation. The book examines evolving ways in which these are being managed. Again the theme of increasingly abstract skills appears. But there is a limit to systemisation. Inevitably systems will require very disciplined work-practices by those who create or run them. But none of this is new - it is only the continuing evolution of a society which has solved these problems before. But like driving an accelerating car, to keep control requires more concentration and more understanding of the road ahead. This book does not give any maps - it does tell you about the beast you are driving.

The evolution of systems and consequent momentum towards abstraction is a perspective that could only have been reached towards the end of the twentieth century. It would have made no sense to the inventors of the great manufacturing, trading and governmental systems which have transformed society. This perspective does not have to delve into technicalities to explain it. The historical and social aspect of systems is independent of an exposition of the underlying logic which was never required by the pioneers of the process.

The ideas behind this book had their first glimmerings when I outlined a course for technology students which would give those students a social perspective to which they could relate. This was prompted by the tales of dismal reactions to attempts to "humanise" engineers with compulsory liberal studies courses. Some years later I found myself having to prepare a course for information technology students aspiring to leap into management. This course, part of the Graduate Diploma in Information Management at UNITEC Institute of Technology in New Zealand, provided the impetus to write things down. Crucial ideas for the lecture series were crystallised during a conversation with Isobel Hawley. This crystallised many of the ideas outlined above. The book has dispensed with the technical content of the course (which legitimised it for its intended audience) and has included a great deal more. Many people have contributed to the journey from lecturer's notes to a book for the general reader. Joan Bain heroically proof-read an early version of the course notes. I am grateful to students who suffered and struggled through the course.

Students, like patients, are always a source of tests and refinement of ideas. Polly Joseph convinced me that the book fills a niche for the general reader and so caused some restructuring of the text. Rachael Macfarlane and Margaret Kennedy provided criticism of a first draft which prompted further revisions and improvements. L.R.B. (Bob) Mann and Cliff Nighy both read an early draft and gave useful advice and much encouragement. Cliff Nighy gave the final draft a valuable scrutiny providing advice and corrections. Finally Michael J. Pomerenke provided a much needed proof-reading of the manuscript. Nevertheless the responsibility for remaining errors lies with the author. Finally the book owes much to my wife Kwai Ying for prompting the circumstances to write a book - not just a set of lectures.

Chapter 1: Systemisation, Abundance and Competition

The greatest transformation in the daily life of the common person has taken place since the late nineteenth century. Where physical toil and drudgery was the lot of most people, the developed world now has a glut of food, more goods than a nation can buy and shortage of exercise. The key to this change is not so much science as the development of manufacturing and farming systems. Requiring less than a quarter of all employees these systems supply more than we need.

Prelude to Systems: Standardisation

All history is full of twists and turns. We are in the midst of one of the greatest trends in history which had an important impetus in the failure of the first mechanical computer. In the 1820s, Charles Babbage had managed to obtain a fabulous sum of what we would now call venture capital to build a machine which could recalculate logarithms. Tables of logarithms, known to generations of school children up to the 1970s, were the basis of navigation. Errors in logarithms, of which there were many, meant errors in navigation. Errors in navigation could mean death, insurance claims, unwanted expenses at the British Admiralty. £20,000 in the 1820s was cheap for a device to correct the tables (and win back some prestige; as the French had the best tables at the time). Babbage proposed a "Difference Engine" which would accurately calculate logarithms to 20 decimal places. Babbage had already produced an impressive calculating device - a good start to obtaining the government support. The Difference Engine was a much more ambitious device and would require thousands of identical components made to precise standards, something that was difficult to accomplish at the time. The problem was to get cogs, wheels,

1

axles and so on to be made to high precision. Cogs and wheels had to be able to turn thousands of times in a single long calculation. At the end of this, their positions had to be precisely determined. If not errors would occur. One craftsman, using one set of measuring tools, could produce all the pieces for the machine - the whole forty thousand. This would be unlikely. If the work were allocated to many craftsmen, each with their own measuring tools, they would all have to be calibrated to the one standard, and the copies would have to be highly accurate. The deviations from specification would be numerous and controlling them would be nearly impossible in the circumstances.

Babbage collaborated with Joseph Whitworth who would later invent the standard Whitworth bolt. But the lack of accurate, precise standards hampered the birth of the Difference Engine. The engineering problems made progress slow. Babbage came up with a new design for a machine called the Analytical Engine, which was capable of storing both programs and data. Again, engineering problems, the lack of components conforming with a precise standard, defeated the project. As fate would have it, the Difference Engine also served as an early example of Britain failing to capitalise on its ideas. Babbage's ideas circulated in Europe, and eventually in the 1850s, G. Scheutz, a rich Swede with better access to precision engineering than Babbage had had 20 years earlier, made a Difference Engine. It won a medal at the Paris Exhibition in 1855. A second Difference Engine was built in 1991 by the London Science Museum and was found to be capable of evaluating seventh degree polynomials to 30 decimal places.

Whitworth was the more successful one of the collaboration. His name is associated with precision engineering. He had worked in the workshop of Henry Maudslay who had invented the metal lathe in the early 1800s and had perfected a measuring device accurate to one ten thousandth of an inch. Whitworth carried on this work. In 1856 he produced a machine which could measure to an accuracy of one hundred thousandth of an inch. He produced a simple bolt, but he produced them in huge quantities. It was not necessarily the best thread, but all the bolts were the same. The Whitworth indexing machine could produce the same bolt everywhere. They swept the market. It was an early application of standardisation of components. He wasn't the first to build standardised components but he made very significant advances in the precision of standard or interchangeable parts.

Eli Whitney was an early American pioneer in using standard components, having established a gun manufacturing plant in Connecticut by 1782. This used stocks of pre-made parts for the assembly of guns. But guns are made from small components which can come from individual workshops. The key to the industrial system comes from the standardisation of large components. The use of dies, stamping and pressing was a crucial development. It is one thing to stamp out coins and quite another to stamp out metal lifeboat halves. This was accomplished by 1850 with a hydraulic press which could produce a pressure of eight hundred tons. It also made possible the now ubiquitous farming material, corrugated iron. The hydraulic press, which was invented by Joseph Bramah in 1795, must be counted one of the most important inventions of all time. Bramah was a cabinet-maker who invented a complex unpickable lock which he decided to manufacture. He hired a young blacksmith to help him make the lock. The blacksmith was Henry Maudslay. Maudslay invented the crucial leather seal used in the original hydraulic press. Bramah was a prolific inventor, having to his credit the flushing water closet and hydraulic transmission power. Maudslay was later a partner of Mark Isambard Brunel who invented a revolutionary tunnelling shield in 1818 which was instrumental in the eventual completion of the Thames tunnel. Maudslay continued as a brilliant inventor and another Maudslay protégé, Alexander Nasmyth , patented the steam hammer in 1842, so adding to the ability to forge standard products. This line of brilliant creative inventors contributed enormously to standardisation and precision engineering and helped lay the foundations for a huge increase in productivity and the creation of wealth.

The impetus for standards came from many directions. Insurers want standards so they can build statistics on the likelihood of having to pay-out. At one stage boiler explosions were killing 70 to 100 people per year in Britain which forced manufacturers and insurers to combine to impose standards on boiler manufacturers.

If measurement in engineering is important, it is crucial in products from the chemical industry. If dyes are not fast or pharmaceuticals not mixed to high degrees of accuracy, disaster can follow. Justus von Liebig (1803 - 1873) in Germany, stimulated the move to standard chemical apparatus and methods of analysis and so helped to initiate the rise of the German dye-stuff industry, agricultural science and, ultimately, agribusiness. Standardisation, the great breakthrough of the nineteenth century, was breaking out all over.

The Analytical Engine, which embodied all the principles of a modern computer, relied on the clever idea of J.M. Jacquard of coding weaving patterns in an array of holes in cards. A series of metal probes was pushed against a card by small springs. If there was a hole, a probe would go into it and be in a position to be lifted by a rod. The probe was connected to wire holding an eye through which ran the weaving threads. Only those probes which went into holes were lifted when the rod moved up and these in turn lifted the weaving threads. Jacquard's loom, invented in 1801, was an immediate and enduring success. It is very frequently cited as a computing machine with the special endearing characteristic that it used punch cards. However this ancestor of electronic computers is simply a clever, easy way of triggering mechanical events and has no special characteristics beyond that it was a way of distinguishing one physical configuration from another. But then that is information.

Babbage's Analytical Engine was probably the most sophisticated system of its time while the Jacquard Loom was the most widespread and useful. "System" is a vague word but here it means something which takes clearly described inputs to which it applies a standard transformation in order to produce outputs in a reliable and predictable way. The pioneers in social systems were the Chinese in their "hydraulic society". In order to control rivers and prevent flooding, the Chinese produced a civil service which was to be a system. Standard inputs, assured by preliminary testing, went through (more or less) standard instruction to emerge as "trained" outputs. This was the first large-scale attempt to produce an education system with a certifiable output - the civil service had to pass exams. The certified output was then of a standard to carry out duties in the empire. We now have legal systems - same crime in the same circumstances transforms to the same punishment. Welfare systems - same need transforms to same welfare support.

Systems are boringly impartial and frequently infuriatingly indifferent to human variations, especially when we are on the (possibly not) receiving end. Boring and infuriating, they are a society's only bulwark against caprice, chance and partiality. Much of what we think of as civilised is the result of the creation of social systems. Most of what we now accept as prosperity is the result of the creation of manufacturing and distribution systems. This creation of systems of production and distribution is *systemisation* : the replacement of rules of thumb, ritual and hereditary trade secrets with the creation of documented rules and functions designed to fulfil some overall goal of an organisation. The definition, such as one sentence can capture it, obscures a long evolution in thinking and one which is far from complete.

Humans are good engineers. Thinking about and then making things, the core of engineering, along with language, is what has distinguished humans from animals. Particularly the grand scale on which we indulge and mix these two skills. The spear, the knife and the bow and arrow are the crucial inventions of human evolution. It comes naturally to us to lash things together, to use the springiness of wood and to counterbalance or wedge material to form shelters and bridges. We easily understand the forces of air and water. All these things were available to humans for thousands of years before the tentative mathematical and physical descriptions were attempted. Handling the physical object is all the explanation we need. The physics, the diagrams, the explanations are not natural. It still takes some effort to describe a watch or a combination lock without a diagram. It takes more effort to describe them in terms useful to a computer.

The obviousness of a mechanical device becomes a long, forced description in terms of "whenever such and such", "while this part is doing...", "if the ... then.." which are so familiar to computer programmers. It can be done. Carburettors, anti-lock brake systems, governors and mechanical controllers of all sorts, even aircraft autopilots, have yielded to computer description. The programs are inevitably difficult to fathom in comparison to the obviousness of the physical system. The computer description has now been extended quite successfully to the less obvious chemical and industrial biological systems. But there is an underlying reason why this can be done. The physical, biological and chemical systems of interest are built as systems - to reliably convert one set of inputs into something new. It is this reason that permits the computer description. The devices are systems: they are there to work as mechanical, chemical or biological functions, transforming forces, chemicals or biological inputs into desired states, chemicals or biological products.

The term "mechanisation" has been used to describe the march of inventions through the industrial revolution to our time. But "systems" is a more general term and captures more of the ideas which are behind the devices which now enrich us. It is also closely linked with the concepts of information. In a system, "information" refers to anything that can be used to distinguish between one thing, or state of affairs, and another. It permits the systems to distinguish between configurations and states such as valves or electrical circuits being open or closed, levers and bars being lifted or in place, the concentration of chemicals, or whether nutrient is flowing. With this information things can be changed, processes can be started or stopped. And, of course, it all works "like clockwork."

Until the computer, the paradigm for a system was the mechanical clock. This appeared in Europe from the start of the fourteenth century. Until then the most complex astronomical clock had been built in 1090 by Su Sung and collaborators in the Chinese capital K'ai-feng. This device, the size of a two storey house, was based on a water wheel. The most renowned early mechanical clock was built by Jacopi and Giovanni di Dondi. This device, an *astrarium,* was started about 1348 and completed and documented (130,000 words) by 1364. It was without precedent in its complexity, precision and documentation. It told the time and displayed the positions of the signs of the zodiac, the five known planets, the sun and the phases of the moon. It used the technique of verge and foliot escapement fitted with a mechanical weight. This technique was invented by an unknown genius about the end of the 1200s. Each clock was a work of mechanical art produced by a team of craftsmen.

The achievement of the Dondis is enormous. To see the possible applications of the simple verge and foliot technique and put it all together is remarkable - given that there were no examples of similar devices. It is hard to think of another such innovation so rapidly developed. Perhaps Babbage's computers, had they been brought to fruition by their inventor, would have provided another example. The combustion engine and the car, taken from Benz and Daimler's first machines to their mass production by the First World War, is perhaps the only comparable technology.

The clock as a system transforms equal intervals of time, as measured by the movement of the sun or stars, into equal differences in the change of the displays. Socially it transformed public life by making precise appointments possible. It became possible to be late or early. It forced social intercourse into the boundaries of hours as designated by the movement of mechanical weights. Time had become *minutely* standardised as had length and weight. This opened the way to the mathematical expression and testing of physical concepts.

The late Middle Ages were powered by wind and water. The French historian Jean Gimpel contends that the first industrial revolution occurred in medieval times, roughly 1100 to 1380 when the catastrophe of the plague slowed the march of industrialisation. To make a flour mill, or to pound fibre using water or wind requires transferring the primary source of energy to where it will do work. This required thinking about chains, belts, cogs, axles, cams, toothed wheels and so on. It also required taking into account the performance of materials, including those being worked. The entire enterprise is designed to produce reliable actions from a steady, if not standardised, input. The beginning of systems thinking.

What is generally taken to be *the* industrial revolution started with the development of the steam engine piston which allowed the transformation of heat to mechanical energy. The invention is credited to Thomas Newcomen about 1712. This was a very successful machine, easy to build, reliable (by the standards of the time), and adaptable. James Watt added the crucial improvement of the condenser to it. Watt had set himself up as an "Instrument Maker to the University College of Glasgow" which included among its few teachers the philosopher Adam Smith and scientist Joseph Black who was a pioneer in the study of heat. He was called in to repair the College's scale model of Newcomen's steam engine. Although Watt learned something of the prevailing ideas about heat from Black, there is no evidence that Watt used any of Black's ideas. Watt was unusually talented at mathematics and had made himself familiar with the works of Newton, some of whose observations confirmed Watt's own investigations of heat. Watt patented his machine in 1769. The first venture to produce Watt's improved steam engines failed, bankrupting his partner John Roebuck. Eventually, with the help of the manufacturer Matthew Boulton,

Watt was to successfully produce the machine. This invention allowed energy locked up in anything that would burn, to be converted into mechanical power, well away from the fuel. This freed sources of mechanical power from local sources of water or wind power. A systems approach was required to distribute the power around a factory. If the steam engine gave a steady supply of power then the problem was to distribute that through as many productive devices as possible. This led to the understanding of the conversion of heat to mechanical energy. But more of that later.

James Watt did not merely refine Newcomen's invention, he established a factory to build the engine. The task of building an engine was broken into standard operations and the factory arranged for maximum production. Tasks and components were standardised as far as possible and workers paid piece rates. Watt's insight into the systemisation of production is probably as great a contribution to the industrial revolution as the refinements he introduced to Newcomen's pump.

A number of important inventions became grist to the systemisation mill although they were not always as influential as they should have been. The assembly line, as a moving conveyor of items which are added to and which transformed materials to obtain a final product, was originally used for grain milling and then fifty years later to make biscuits (but not from the same flour). Oliver Evans in the United States invented a flour and grain mill that required no added human labour. This was in 1783 and a patent was obtained for it in 1790. But it was before its time. It was judged on its parts not on the whole system. And this is the fate of systems. Individual parts need excite no one. The connections between the many parts are the things of interest. Thomas Jefferson commented on the ancient origin of the parts of Evans' invention: "The elevator was nothing more than the old Persian wheel of Egypt, the conveyor is the same thing as the screw of Archimedes".

In fabric manufacture, the prize for the earliest fully automated silk loom goes to Jacques deVaucanson (1709-82). He was a system builder *par excellence*. His reputation was made with automata, those mechanical devices which played tunes or mimicked animals. He was a master of detail. His flute player had lips which moved and a tongue which acted as an airflow valve. His mechanical duck not only quacked, flapped, swam and waddled, it picked up grains, swallowed, digested and excreted them, so laying claim to being the first miniaturised chemical digester. By 1756 de Vaucanson had a silk factory with mechanical looms, a full two decades before Richard Arkwright established Manchester as the centre of cotton spinning using James Hargraves' "Spinning Jenny" looms which were invented in 1767. De Vaucanson's factory, like his automata, was planned in detail. The attention to lighting and ventilation anticipated modern factory design by about two centuries. But in spite of this chance to dominate the textile trade, France was not quite ready. The vapid aristocracy didn't care. It was only after the revolution that Jacquard was able to make his loom a commercial success. The success of inventions requires the right social conditions.

Textiles and clothing have been at the forefront of manufacturing since the Middle Ages when wind and water mills were used for "fulling" cloth. This is pounding cloth in water to clean it before it is sold. The weaving looms invented at the start of the industrial revolution were a crucial step in the development of manufacturing systems. However progress did not stop there. Isaac Singer's foot powered sewing machine introduced in the 1850s was not only a consumer item but an important factory device. The speed of sewing machines is indicative of the increase of productivity in the manufacture of clothing. This rose from about 20 stitches per minute in the middle of the nineteenth century to ten times that amount at the end of that century. By 1970 it had reached 7,000 stitches per minute. Stitching is not the only aspect of productivity. In the 1950s, clothing factories started to use automated fabric spreaders and cutters. Automated computer controlled cutting was introduced in the late 1960s. The use of computers to work out cutting instructions which minimised waste was an early use of optimisation in manufacturing.

The nineteenth century produced the understanding of mechanical and then electrical systems. Understanding and application were seldom separate. Each application was driven by new ideas for manufacturing which produced a surge of productivity which, in turn, forced innovations in factory organisation and the co-ordination of materials and labour. Co-ordination required information and the telegraph, and later the telephone, enabled the rapid transmission of data; the start of information technology preceded computers by about a century.

Although international trade has existed for much of recorded history, by the end of the nineteenth century the shipment of goods overland grew enormously with the spread of railways. The steam engine, with its application to rail and later sea transport, was the great new technology of the nineteenth century. The railways themselves were made possible by the standardisation of tools proceeding from Whitworth and his American counterparts. They also provided a huge market for steel and ushered in the Carnegie steel empire. A number of railway collisions spurred the urgent requirement for co-ordination. The telegraph, commercialised in 1840, provided the networking tool to co-ordinate railways. This would not have been successful without prior improvements in steel making and drawing and wiremaking which derived from improvements make by Henry Bessemer .

The railway corporations were among the first to face and solve problems of integration, of running trains back and forward on single lines and co-ordinating through timetables, signalling and the telegraph. Learning from this, the large manufacturing corporations in the United States grew by harnessing standardised methods for production, the logistics advantages of rail transport, and the communications technologies of the telephone and telegraph to co-ordinate supplies and sales. The age of integration had begun. Through standard inputs, processes and training programs, standard products could be produced and sold. The Empires of Rockefeller, Heinz, Eastman-Kodak, Dupont, Carnegie and numerous others required integration and co-ordination. They required the civil army of paper-workers.

The rise of these companies has been studied by Alfred Chandler who argues that a crucial advantage was gained, again and again, by producing a large number of products from a small range of feed stocks transformed by variations of a small number of core processes. The processes were simplified so that hunches, long experience with the materials and complex thinking - all of which slow things down - were eliminated from the production processes. This ability to produce many things from few inputs is what Chandler calls the scope of the enterprise. The scale is the quantity that could be processed. The final ingredient to the corporate mix was the professional manager. Accountable managers were chosen, not on the basis of being known to, or related to, the owners, but on the basis of demonstrated (or claimed) expertise.

The success of this recipe, along with the usual dollop of luck, is demonstrated by the rapid rise of the cigarette business. Cigarettes were introduced to Europe from Turkey after the Crimean War. In 1880, James Bonsack invented a machine which could automate their manufacture. American Tobacco bought the device and by 1883 could produce a billion cigarettes a year. The rate of production could easily outpace the desires of the local market. American Tobacco merged with British Tobacco to gain access to the European market and the British supply of tobacco from Kenya. This is a classic story of *horizontal integration*; gaining access to larger markets through joining, merging or taking over other firms in order to gain their customers. The similar process of merging or taking over companies in order to gain control of supplies is called *vertical integration.* Vertical integration is, above all, a tool for the standardisation of supplied goods. It can also absorb the supplying firm into the administrative system of the manufacturer.

Manufacturing corporations appeared very quickly in the United States - effectively in one generation. In 1878 the New York Stock Exchange traded the stocks of railways, coal mines, telegraph companies, express companies, a land company and a steamship company. There were no public manufacturing companies. Much of the steel for the railways was imported. By 1900 all this had changed. The stock of manufacturing companies dominated the exchange. The United States no longer imported steel. Carnegie's steel companies, on their own, out-produced the entire British steel industry.

The change to large-scale manufacturing was not without its pioneering intellectuals and consultants. The polymath Frederick W. Taylor called himself a "consulting engineer in management" and was a very successful prophet of scientific management. This was the idea that work could be made much more productive if it could be broken down into short, quickly executed tasks. His name is linked with the much resented time and motion ideas of management. But such ideas sprang from the impetus to improve productivity and eliminate waste by standardising tasks and processes. Just as Adam Smith was much more thoughtful than his image as an unbridled free marketeer, so the truth about Taylor contains more than the 'man with the stopwatch'. In his study of the company man ethos, Anthony Sampson writes, "In fact Taylor was much more humane than most manufacturing bosses, and he was much concerned with harmony in the workplace. In his classic *Principles of Scientific Management*, published in 1911, he argued that workers and bosses should have more equal responsibilities, and that workers should be encouraged to suggest improvements - which today sounds very Japanese. He insisted that managers - a word he disliked - must study the workers' characteristics to help them advance, and become servants of workers, not their masters. Scientific Management, he told Congress in 1912, 'is not holding a stopwatch to a man'. He called for a complete mental revolution 'which must be equally complete on the part of those on management's side'. His champion today, Peter Drucker, claims that Taylor 'had as much impact as Marx or Freud'."

Taylor in steel and engineering, Carnegie using the inventions of Bessemer in steel, Rockefeller in oil, and Heinz in canning, pioneered many of the new techniques of integration and standardisation and created the momentum towards large-scale manufacturing systems. These systems found their classic expression in the work of Henry Ford. The breadth of his vision combined with the ability to make it happen make this man the genius of large-scale manufacturing and the man with the greatest effect on the twentieth century. His accomplishments were probably inevitable; without him we would still have large-scale, highly systematic manufacturing organisations. But it would not have happened as fast as it did. While the twentieth century has had its share of insane dictators, brilliant scientists and inventors, the life of the average person, taken over the whole century, has been most affected by the motor car as a consumer item. The success of Ford's Model T as a car for the masses even caused a jump in farm productivity. As farmers adopted motor transport, they brought fields reserved for horses into production. From cheaper, fresher food to the growth of suburbs and new ways of love and leisure, the car, as part of the average package of goods, has changed the material expectations of the twentieth century. Ford had the clearest vision of a mass market built on cheap goods and well paid labour. Ford's initial venture into mass production was the famous Model T. It first left the famous Highland Park factory, in 1908, at the rate of 100 per day. At the end of its life, nineteen years and sixteen million cars later, it could be produced every 24 seconds. Modern plants do not need this level of hectic production. Ford Australia, producing for a saturated market, produces at a much more leisurely rate: 400 cars of a variety of types and models are made each day, or one every 3 minutes 36 seconds for twenty-four hour operations. This is a small plant by world standards. Car manufacturing requires enormous capital investment which can only make sense if the economies of scale are large. This only occurs if the market is large enough to absorb long runs of single products. In the 1980s annual production runs of 200,000 to 400,000 were considered optimum.

Henry Ford rigorously integrated supplies from iron mining and rubber growing. He controlled ships, a coal mine, a glassworks, timberland and associated sawmill. All the manufacturing management techniques were pioneered: just-in-time processes were routine as were materials resource planning and supply chain management. Iron ore from Ford's mines arriving on Ford's ships would be processed at the giant River Rouge plant, into finished cars in 28 hours.

Ford was a cultural phenomenon. He symbolised and legitimised manufacturing and the use of the combustion engine. Tractors became fashionable as automobiles became "domesticated". US tractor numbers went up twenty fold between the world wars, from 80,000 at the end of the First World War to 1.6 million by the start of the second. Agribusiness replaced farming as business concepts took over. By 1926 a Californian agriculturalist would claim, "We no longer raise wheat here, we manufacture it..." The car did, however, have a ruinous effect on rural social structure. Small stores were by-passed as farmers drove to larger centres. The cropland reserved for horses was brought into food production which was in excess of demand, so depressing prices. In 1921 there were 29,788 banks in the United States. Most of these were small banks serving rural communities and providing agricultural loans. Mobility meant that farmers could explore their area for competitive interest rates which set the scene for the bank failures to come at the end of the decade.

Ford did not have car manufacturing all to himself. Indeed there were about 600 car firms that came and went in the early part of the century. The major competitor was his contemporary

William C. Durant who formed General Motors. General Motors weathered a number of financial crises until Alfred Sloan became its president in 1928. Where Ford was a manufacturing genius who hated bureaucracy, regarding managers as little more than components of the system, to be replaced as they wore out, Sloan was a managerial and organisational genius. While Ford pumped out the Model T, Sloan set up five separate divisions to produce a range of cars for various prices. He decentralised production while centralising management, so setting a pattern for all large-scale manufacturing, including, eventually, Ford Motors. This innovation introduced the staff and line organisation with its autonomous divisions and regular reporting through hierarchies. It also allowed specialisation within a company and the ability to sell components to other divisions throughout the world. The now common practice of producing components for rapid assembly in another country - the "knocked down" version of the product - arose from Sloan's organisation of General Motors.

The fortuitous interaction between the scale, the productive power, and the scope, and the ability to produce a large range of goods, determines how competitive an organisation can be. Historically, for a given level of technology and market size, if the optimal plant size was achieved with very significant cost reductions, the market was dominated. Market domination ensured the flow of money to keep ahead of competitors by researching and deploying new technology.

This is still true today, although plant size is replaced with organisational integration and alliances of expertise.

It was not just machinery which was important in the development of the industrial system. The crucial business concepts in commercial law, accounting and investment developed rapidly throughout the nineteenth century. These had a long evolution. Modern accountancy dates from Paciolo's treatise on accounting and book-keeping published in 1494. This was really a compilation and sifting of practices which had been adopted by the Italian merchants and bankers of the time. The Dutch, more than anyone else, refined and developed banking and investment in the 1600s. The famous speculation in tulips in the 1630s was the first speculation boom - and bust. The public share, along with the limited liability company, are wonderful devices for moving money to where it can be applied to create future wealth. Stock trading and banking were also extremely well developed in the United States by the First World War. United States money was to help Britain and France grind away at Germany so that after that war Europe was exhausted and the United States the pre-eminent world power. In the Boer war, United States banks, pre-eminently the Morgan Bank of J Pierpont Morgan, had lent Britain over 100 million dollars. During the First World War it would organise loans to the tune of $1.5 billion which was spent at the rate of 5 million pounds per day. A great deal of this flowed straight back into American industry. Bethlehem Steel took contracts of over $US135 million to provide arms to Britain. Du Pont, by supplying 40% of the allies munitions, evolved from a gunpowder company to the world's largest chemical producer.

The historian Richard Overy has argued that the Second World War was as much a clash of production systems as armies and ideologies. The great developments of standardisation and manufacturing systems in the United States were the ultimate weapon in that war. Brilliant engineers on both sides of the war produced high-performance aircraft, tanks, guns and ammunition. In the end, as men and machines were destroyed, it was the ability to replenish armed forces quickly that made much of the difference. Individual weapons such as the Mustang fighter and the Soviet T-34 tank played decisive roles but they had to be available in sufficient numbers The advantages of long-run production systems and standardisation gave the advantage to those who best mastered them. The Nazi system hampered the German industries by putting them at the mercy of Nazi hacks who asked for constant improvements and innovations which resulted in short runs of high quality machinery. The number of products was astonishingly high: at one stage there were 425 different types or variants of aircraft, 151 types of lorries and 150 types of motorbikes - a nightmare for maintenance.

The Soviets adapted, as much as they could in war time conditions, the American techniques of large-scale production of standardised but modifiable products. The most decisive of these was the T-34 tank which was made in the huge tank city "Tankograd". As production processes were improved, stamping rather than casting of tank turrets was used, as was automatic welding of parts. Whenever possible, simplification and standardisation were used to aid automation. It was these tanks which turned the tide of the war during the greatest tank battle ever at Kursk where 2,700 tanks, 10,000 guns and 2,000 aircraft on the invaders' side faced 3,444 tanks, 19,000 guns and 2,900 aircraft of the Soviet Union. The ensuing battle destroyed so much Nazi ordinance that the German army was from then on progressively "demodernised". Amongst the heroes must be counted the mainly women workers who worked fanatically, in appalling factory conditions and on the brink of starvation, to create the tanks.

In the United States, mass production was driven to even higher levels. Exceeding all expectations, the United States switched its predominantly civilian economy into military production so quickly that by 1942 it produced more military weapons than the combined Axis states. It out-produced the Axis by 20,000 more aircraft (170%), 13,000 tanks (218%), and six times as many guns. Mass production reached new heights. Henry Ford brought his irascible genius to the production of the B24 Liberator bombers , creating the mile long "Willow Run" plant. By 1943 this produced 10 bombers a day and during 1944 the promised rate of a bomber an hour was reached. In shipbuilding, Henry J. Kaiser master-minded the mass production of cargo ships - the Liberty Ships. It took 355 days to build the first one. Eventually this was cut to eight days.

This history renders puny the posturing of information technology to be the vanguard of productivity.

The Liberator required thirty thousand different parts and required one and half million parts for the total effort of assembling the wherewithal to put it all together. To produce such prodigious quantities of complex machinery requires a virtuosity of integration and a mastery of systemisation. In order for things to go smoothly, rules must be created. The staff of distant supplying or distribution firms cannot be allowed to do things on their whim. Judgements, such as deals with suppliers or customers, must be justified. Rules must constrain the interactions of staff in all companies concerned. Documents must conform to certain requirements so that time is not wasted searching for additional information. Processes must be carried out in conformity to rules or policies. Staff must be treated according to policies. The great vertically and horizontally integrated machine is also a bureaucracy. Orders in one day, delivered somewhat later, in good condition or else your money back. Inevitably, this translates to a desire for further automation and systems. Automation of processes produces greater control over processing and, eventually, the desired cost reduction without compromising the quality of the goods. Systems do the same for administration and control. Their creation demands the ability to see or create structure where none previously existed; and to create rules which are sufficiently apposite, and well expressed, not to cause problematic contradictions or restrictions.

The closer the administrative, control, and operational structures can be welded together, the more productive the organisation will be. To implement the vision of a large-scale business, with many interacting divisions and associate companies, required new structures for administration and communication which, in turn, provided a ready market for new business specialities in management and operational technologies. The growth of service organisations is closely allied to the complexity of running large corporations. Engineering, accounting, marketing, computing and human resources services have arisen out of problems of these complex systems.

Perhaps there is no better measure of the extent to which a society has built on the accumulation of techniques than its investment in tools per person. Tools are any pieces of equipment which help a person do a job. They include the equipment used by a hairdresser, a surgeon, an oil prospector or a train driver. They include the cable and computers which deliver cable TV and the sewers used for waste disposal. In short, tools include implements, systems and infrastructure. Using this broad definition of tools, each society accumulates stocks of tools and uses them to create and maintain its wealth. Pre-industrial societies had few tools. Nomads carry their stock with them. The accumulation of systems and the accumulation of tools go hand in hand. Each new tool slots into a system of tools. Concepts are also tools. The imaging tools of modern medicine: CAT scanners, MR Imaging and Ultrasound, require not only tools for producing X-rays, extremely powerful magnetic fields and ultrasonic sound, but also computers to process the signals and present them on a screen. The programs in the computers rely on mathematical tools which have been developed over a period of a hundred years up to their use in these tools. Tool making borrows on everything, and as it accumulates it incorporates unexpected aspects of conceptual, physical and biological tools. In his book *Creating Wealth*, Lester Thurow proposes as his eleventh rule of creating wealth, "Only those interested in the future build tools. Whatever they might say, those who build few tools are not interested in the future." He states that, "In 1997 Americans worked with $24,883 billion worth of tools. Each worker had $191,000 worth of tools. Each $1 of gross domestic product generated, required $3 worth of tools. Business plant and equipment and residential structures each accounted for 35 percent of this total, public tools (e.g. roads, sewers, schools, airports, military equipment) accounted for another 20 percent, and consumer durables (e.g. cars household equipment) the remaining 10 percent."

The Second World War showed the power of a new tool: automated systems for calculations and for checking combinations. The computer married calculation with logical manipulation. Calculation was required for ballistics and was pushed along by the atomic and hydrogen bomb projects. Logical manipulation grew from the British code-breaking work. Within twenty years of the first steps, computers would be widely used in administration, business operations - initially in searching the best way to do something - the control of devices and calculations for the military. The 1950s were the time in which the term "systems analysis" was born. Thus the profession of systems analysis became wedded to information technology, an accidental and probably unfortunate marriage. Information technology is only about systemisation. Engineers create systems but engineering didn't absorb the word "systems". That was handed to computer specialists. The prophet of "scientific management" would have been delighted. If the nineteenth century discovered standards, the twentieth century fulfilled the dreams of Frederick Taylor and discovered systems.

Systemisation: Managing Globally

Investment in factories, production devices and the integration of conglomerates produces formidable production systems. Time and again this recipe has saturated the national markets with a set of commodities. The next step is to "go global".

Trading in many nations, with both customers and suppliers in different countries, is more risky than trading in one's own nation. These risks include risks to property, personnel and plans from political upheavals, and risks from misjudging the motivations of personnel and customers from a different culture who might have, for example, different attitudes to credit, interest, sources of authority and discipline. Included in this group of risks are ethical ones, which arise from attempting to by-pass cultural connections in order to use a person's expertise to advantage. Transport costs, including graft to ensure safe passage, and their attendant insurance problems, can increase costs. Foreign exchange risks can destroy narrow margins on high volume, low margin products.

Three things mitigate these risks. First is the access to new markets. Second is the access to cheap labour, still an advantage but a declining one with increasing automation. Third is the ability of multi-nationals to offset risks by moving money to branches in countries with the most favourable tax regimes, and to do this when the exchange rate is favourable. All of these require a sophisticated level of co-ordination in the marketing, operations and especially the financial systems. Particularly the financial system if the movement of money is to gain the best advantages from the constant fluctuations of exchange rates. This type of capability frees a multi-national company from any country. Asea Brown Boveri was originally a Swedish company which merged with a German company which had already merged with an English company. It has divisions all over the world, its headquarters are in Switzerland and the company language is English. It is a company without nationality. Similarly, Coca Cola has declared itself to be, not an American company, but a World company . "World Class" and "World Company" are the desired accolades of successful companies. The next step in this progress is not only to break the links between company and nation but between company and the manufacturing of a product. Here Nike is the prime example - it is a brand and it designs things associated with the brand. It accepts tenders from those who would produce the branded item. The attraction for the producer is a near guarantee of sales. Nike then collects most of the high profit that comes from the association of the brand. That is all. It has been taken to task for not caring how the branded items come into being - for example in Asian sweatshops.

The sophistication of risk management of these world traders can be seen in new techniques such as "hedging", one of the commonly used tools of financial management. The hedging of foreign exchange against fluctuations is now well established. It represents managing chance. This is no trivial business but one of the largest in the world. The nominal value of risk-swapping contracts rose from less than $US3,000 billion in 1986 to $US17,000 billion in 1991. Some things remain outside the control of an individual corporation. Earthquakes and incompetence are two. Losses due to weather, crop or animal diseases can be insured if a willing insurer can be found. If some way is found to make money from risks, someone will have a system to do it.

As production has overwhelmed demand, the effort required to keep goods and services selling and competitors at bay has rapidly increased. This has produced the rise of the marketing professions, and their consequent attraction to the present generation of students. It is no transient fashion. The engineering concept of signal to noise ratio can be applied to advertising and marketing. It takes more and more cleverness, cheek and wit to attract attention in a crowded field of attention getters. It can be helped by deals where one organisation scratches the back of others. The airlines use frequent-flier points and "fly buys" as techniques for getting customer loyalty. Integrated marketing uses computer systems as a way to offer services which would have been extremely difficult to do without them. The problem of keeping track, world-wide, of frequent flier points, demands systems of considerable complexity, especially when special offers of "bonus points" are the attraction. Often the marketing strategy and operations involving an alliance of many companies in different types of businesses, become so complex that it takes extraordinary determination, leadership and the courage to take visionary career gambles to see such complex projects through to completion.

This is illustrated by the alliance between airlines, motel chains and car rental firms.

Each of these businesses has a separate sales cycle:

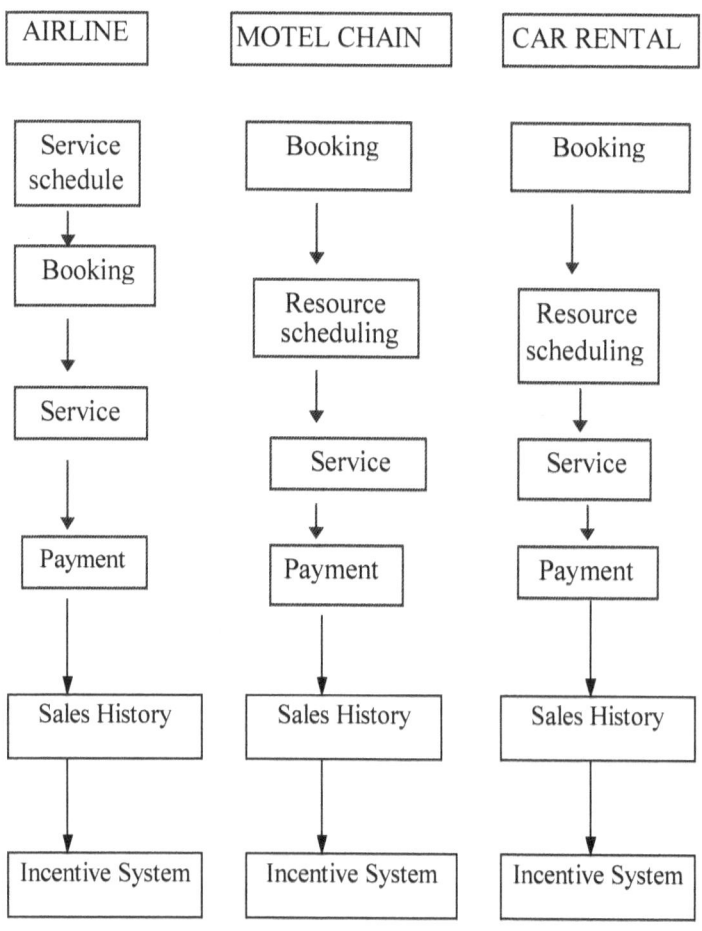

Each one of these systems will be large. There will be a set of options for booking and paying in the airline system, another set for motel booking and paying, and a third set of options in the car rental system. Suppose there are 10 different options in the airline system, and five in each of other two systems. There is no point in integrating all these unless the system can handle all the old options and some new ones as well. There will be at least 250 options that need to be available throughout the system. There might also be a favoured set of options which accrue various types of additional incentives. Thus, the integrated system will be more than the sum of the parts. The ability to cater for so many options takes considerable knowledge and training.

There is a general principle at work which increases the difficulty of large-scale projects beyond the multiplication of choices. The larger and more ambitious the system, the more it must model and prescribe normal business interactions. The business analysts must discover what these are and, almost certainly, the people they interview will not have a clear idea of all the myriad things they do to make the system work. Furthermore, the attempt to capture all the inevitable odd, "exceptional" cases, increases the difficulty to systematise operations. The complexity of the system rises in unexpected ways as analysts render classifications and judgements into lines of logic which capture exploding numbers of options.

The integrated systems can be sketched in the following diagram.

The integrated system transforms three different organisations into one. This can collect and pass on bookings, or receive information on the service "package", bill and reimburse departments so the correct information and money gets to the right profit or cost centre. The system gives many more options and conveniences to the customer. The increased options come at the cost of a more complex system with attendant operational problems.

This expensive exercise locks itself in place by virtue of the effort and investment. It also represents a step in the ratchet of consumer expectations; something that has to be exceeded in the next round of competitive manoeuvres. If the system works well, it increases the "entry cost" of groups of competitors.

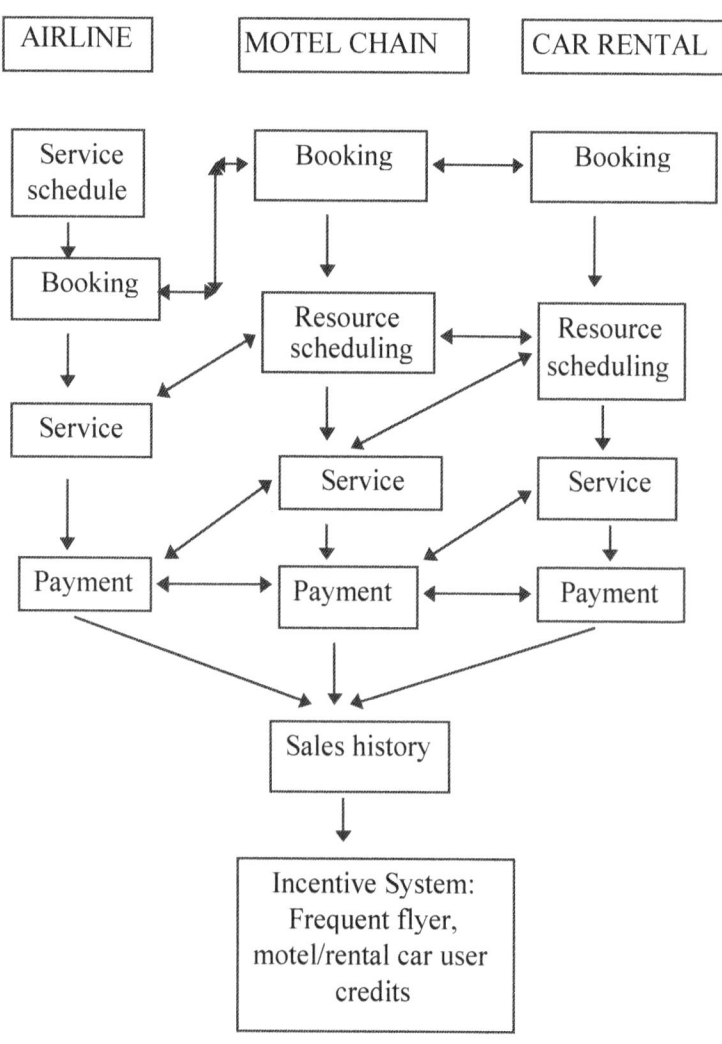

A prime example of how valuable an integrated system can be is that of the Sabre ("Semi-automated business and research environment") system . This computer system, which began in the 1960s as an airline booking system for American Airlines (owned by AMR Corporation), became the tail which wagged the dog. At the end of the 1980s, Sabre was worth about $US 8 billion while the assets of the airline part of American Airlines was worth about five billion dollars. Sabre became a separate public company in 1996. By the end of the 1980s, Sabre handled about 45 percent of all airline reservations made in the United States, 20 percent of reservations for car rentals, and 10 percent of hotel rooms. In 1989 AMR employed 2,400 computer staff and ran two specially configured IBM mainframes linked to seventy-five thousand terminals. In 1999 Sabre corporation introduced "Best Fare Finder", allowing travel agents and consumers to locate cheap "spot" fares. This is a huge co-ordination system which has been evolving for over thirty years.

Of course this is not the only fare selection system. The first system was the Airline Revenue Optimizer developed by Sperry (now Unisys) and Northwest Airlines. This system decides the best price to charge on each seat on each flight, so allowing "real-time" competitive pricing. The system takes into account data of departure, prices of competitors, number of passengers on the flights and so on. Such systems are like privately owned commodity markets. The commodity being airline tickets. This preceded the Internet based specialist commodity markets - such as car parts - by nearly twenty years.

A similar advantage is given to oil companies which have the best geological data and use it to work out the best places to drill. These hard won, expensive data become a strategic asset as they help the company to be among the low cost, efficient producers. So effective are these data, that oil is now easier to find at the start of the twenty-first century than it was for most of the last quarter of the twentieth century when less of it was used up. In the same way, crop producers who use computers to minimise agricultural inputs by a squirt of water here, a puff of insecticide there and a dribble of fertiliser at the right place and the right time, quickly pay off initial investments to become low cost producers.

World companies, alliances, international management, and integrated world-wide systems are now commonplace. The welcoming, even homely facade of franchises, household names and after-sales service are the final customer interface of complex systems of thousands or hundreds of thousands of rules. These represent the experience, and obsessive refinements, of a century of systemisation.

Abundance

Today we live in a world of super-production. Economics might be based on the assumption of scarcity - which in a great deal of the world still holds true - but the real scarcity for business is customers, at least in the (over)developed world. We can, and do, make more than we can wisely consume. The problems of abundant food are obvious. Popular magazines of the developed world carry a steady stream of articles on dieting. Ranked in terms of interest, weight reduction is a significant industry. Entertainment offers so much and so insistently that one wonders how the world got on without night-clubs, shows and numerous television channels. So problematic has the abundance of supply become that countries have instituted laws to increase consumption. The high cost of certifying a car's mechanical soundness in numerous countries is a mechanism for creating the demand for new cars. Planned obsolescence has been with us for several decades. In order to stay in business now, a corporation has to find, defend and hold new consumer niches. This has led management gurus to emphasise innovation, marketing and customer orientation as crucial to the modern corporation. Ceaseless innovation - even apparent innovation - is crucial to keep markets going and prices up. For the most part, developed countries cannot *not* consume. Even small changes in turnover and gross domestic product of one or two percent can be socially disruptive. Any proposal to produce economical cars which would last thirty or forty years is dismissed as being technically unfeasible (which it is not) or as having "no commercial reality" - which is a matter of the way we see things. The lot of the modern corporation is "branding" and marketing alliances, the targeting of the highest quality goods and services in the customers' "price bracket" in order to gain maximum market share or "market penetration", while at the same time searching for new customers to avoid market saturation. All the while producing a stream of apparently innovative variations on the theme.

The problems of productive over-capacity in the large manufacturing corporations which supply the world's cars, trucks, public transport, electrical gear, consumer-durables, aeroplanes and ships, have led to the politics of market access. The General Agreement on Tariffs and Trade, GATT, along with the initiatives by its descendant, the World Trade Organisation, is a gamble by the economically powerful. Their craving for customers - anywhere in the world - exceeds their fear of international competition. This recognises that it is too easy to saturate national markets, that modern production techniques can satisfy markets of hundreds of millions. GATT is not designed to increase world-wide competition. It is designed so the largest corporations have a political agreement to add force to arguments for entering markets in other countries. Market access is usually granted with a deal to locate production in the accessed country. This is a good deal if the wages are lower than in the corporation's existing base of producing countries. The consequence is to drive down the cost of wages and increase world-wide competition. The flood of goods is intensified and, as low-wage countries learn the secrets of production and marketing, existing market niches are made vulnerable to attack.

The dynamics of over-capacity, international wage arbitrage, and market saturation, mean that the way to produce new markets is by new goods which have some claimed superiority over the "previous generation". Companies with any prospect of long-term survival must innovate. The problem of innovation rapidly becomes the problem of funding research and development, which in science-based industries, can become so expensive as to require firms to enter into product development alliances. For example, at the end of the 1980s, Hewlett Packard spent $250 million developing the quiet printer and Gillette, the US manufacturer of razor blades, pioneered a shaver with blades a millimetre wide - a sixth of the width of the conventional blades at that time. "Sensor", which cost $150 million to develop, used moulded plastic springs that allow the blades to follow closely the shape of a man's face. Even almost imperceptible improvements in the lot of humanity now cost a fortune.

The statements of the last paragraph are automatically contentious for economists. The fact that not everyone has ready cash to buy a new car or television set could be taken as evidence for the price being too high. Further competition will bring the price down. Where does such an argument end? When the price is zero? Or a new car is $10.00? What wages are then paid to automotive workers? Perhaps the cars are made entirely in automated factories. How then are the wages earned and the profits distributed back to society? How are dividends for shareholders to be generated? This last question puts a minimum on prices. In a society in which production is automated, shareholders, even if they are large investing corporations working for larger numbers of customer investors, all expect some return on investment - some cut of the profits. This means that there must be profits.

The basis of manufacturing segments of the economy is equations of exchange. These value things in terms of wages which were originally linked to food. The basis for financial models of the operations of any manufacturing organisation is the equation:

Value of Goods = Cost of raw material + cost of labour + energy costs + equipment costs + depreciation.

Each part of the right hand side of the equation can be broken down to have its own equation. Eventually a price must be struck in terms of human needs. A society of slaves is not a good marketplace, something that Henry Ford saw very clearly when he gave his staff the highest wages of any factory workers in the United States. For agricultural products, wood products and minerals, the cost of raw materials can be related to the socially perceived cost of land. The ability to break down the cost of components is the reason why the analytic method has been so successful in information systems and computing. Both accountancy and computing are inherently "recursive": everything to be calculated can be partitioned repeatedly until an agreed set of starting points for the calculation is found. This is seen in the breaking down of costs to the smallest purchased item.

The minimum for the cost of production can be calculated in terms of embodied energy - the total energy required to collect and transform the raw materials into useful goods, including the energy required to build the factory and feed workers. These costs cannot be zero. There will be a minimum. There can be trade-offs between people and machines but there will always be people and there are certainly going to be shareholders for a long time to come. The price of an item then becomes the sum of the actual manufacturing cost, the marketing costs, the targeted innovation costs, the interest on loans and the intended dividend pay-out.

The concepts of embodied energy determining the minimum cost are harder to apply to services as expertise is difficult to cost. For information products that can be easily replicated, the embodied energy is small. Software distribution costs are those of blank CDs, the initial cost of the equipment, and the cost of expertise, which might be in royalties - if they are a consideration - and the rest is profit.

Manufacturing costs include all kinds of infrastructure costs as well. The costs are partly historical, reflecting the way an economy has been "assembled". Modern economies can be seen as a hierarchy of industries. Those at the lower levels producing goods which are desired before those at the upper level. Each level requires that underlying levels are "reasonably active" if not actually flourishing.

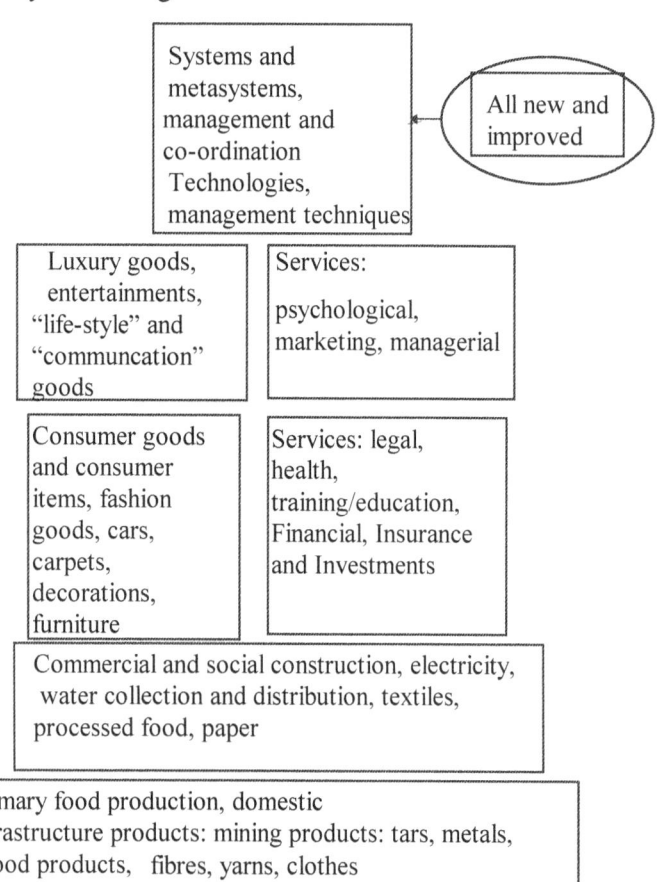

Although systemisation grew rapidly throughout the twentieth century with the top level and the trade in factory systems themselves occurring all though the century, the trade in computer systems started in the 1960s. This trade made new factory designs, control systems and administrative systems into commodities. The next step will be in systems generators and full organisational design systems which will generate the required subsystems. These might be called *meta-systems* - systems which produce systems.

At the beginning of the twentieth century, farming employed roughly one worker in three. Farming now employs two workers in every hundred. Manufacturing rose to include about a third of the workforce but is now around one fifth. The rest serve in the vaguely defined service sector. As machinery and systems made farming much more productive, so automation and systems make manufacturing more productive. Effectively, the wealth of the developed nations issues from a quarter of its workers. And this proportion continues to fall. The rest, in professions, entertainment, education, government, finance, health and tourism, live off this cornucopia.

The top layer of the modern economy is also the sphere of the management of innovation. Although this segment of organisational products is intimately connected with legal, financial, managerial and entertainment activities, it is something new and different. Its effects include the reshaping of the form, if not the content, of many social functions. One particular effect, mentioned in the previous section, is the de-coupling of business from nationhood. Such a move will reshape concepts of for what, or for whom, a person works, and to what entity loyalties are due. (This might well provide a repeat of the church versus state ideological battles. One can imagine a three way split, church, state and business. However, the first two players will probably assume the characteristics of the third so completely that they will merely append the name *Corp,* or *Inc* to their names so that no one will know the difference.)

Value depends on some human desire or need. Food, clothing and shelter are the obvious ones. This is the lowest level of the economy. Most other products have to do with, and in times of over-abundance sold via, desire for security from other humans, for reasons of status and male pride, female vanity and female insecurity. There are, of course, products designed to produce novel experiences and entertainments. The market for services requires at least some pockets where the economy is thriving. Commercial services are found in management and engineering in the lower levels of the economy. But mostly they thrive in the upper reaches. They are most highly valued if they oil the wheels of the economy. So marketing, management, financial, production, system and legal expertise are well paid. Doctors are well paid out of insecurity. The ability of expertise to command a high price has previously depended on some degree of social contribution. Now the association with any known goods, any clear socially accepted item or service, is becoming so diffuse that peoples' names disappear in the "credits" for a product or service (describe your job again?). As more and more people become involved in the minutiae of thousands of entertainment productions, for hundreds of cable delivered entertainment events, status might well accrue to those lucky enough to work at producing physical goods.

The ability to identify one's job in terms of a social contribution is threatened by the general over-capacity of manufacturing and services. This leads to contract jobs rather than permanent jobs and to very high levels of specialisation. (What? You are a Quantitative Analyst for Investment in Nanotechnology commodity futures?) The high level of specialisation is a consequence of systemisation. Human skills and expertise are broken down from a set of heuristics, the rules of thumb which become second nature, to a set of functions. This has happened for centuries. Roman and Japanese sword makers went through complex rituals to produce some of the most impressive products in metallurgy. Now, one factory could probably mass produce all of these swords in a couple of weeks. The mastery of metallurgy is such that most, if not every part of the skill, could be automated. Skills themselves are subject to depreciation - sometimes overnight. Vacuum tube technology, biplane design, and slide-rule manufacture are no longer career options. Systemisation has accelerated the transformation of expertise into a commodity. Each individual has to trade up their expertise as they might their car.

The consequence of abundance of goods and services is competition for customers. This drives down costs for the consumer. Few of us are consumers only. Most of us work in a business. This competition can be fun but it can also be long hours, constant threats of loss of business or employment and, sometimes, unseemly obeisance to the customer (who is always right even if the fool is the Universe's last resting place of good judgement).

In order to bring the cost of goods down, every cost component must be driven down. The one on which systemisation works is the cost of labour. Expertise must be turned into a series of routines, checks and procedures so as to contain and eventually automate it. In terms of administration, this idea is taken to the limit when all activities which can be decided by a set of rules are automated. Staff are then free to do those things which require understanding of humans and social nuances which cannot be included in rules. The only people of value are those who generate ideas and solve non-routine problems.

The result is an organisation of people dealing with customers and making decisions about marketing, investing, planning, scheming and deciding what coffee is best in the staff room. Competition and systemisation produce companies of salespeople, decision makers and "loaders". The last group loads raw materials and data into machinery, including computers, stocks shelves and containers, or they take things out of containers and give them to customers. They are seen in petrol stations, supermarkets, fast food outlets and any type of organisation where personal service has given way to rapid service.

In essence, expertise and, to even a greater extent, information, have arbitrary value based only on marketing consideration. This of course depends on what value society decides to give that product. More than that, most employment and a great variety of products have, and increasingly will have, arbitrary value. This is the future economy. It is a fashion economy and an economy of surpluses.

All this is part of the changes forced by extreme levels of competition. To compete, organisations (routinely) reform company structure, usually with a new, and, of course, improved system. The terms "re-engineering" and "restructuring" are processes of systemisation. The whole company becomes a system to be redeveloped. Historical arrangements and hierarchies are examined for their contribution to getting the companies' products to the customer. This process, which has been going on since the middle of the nineteenth century, arranges the talents of the company in the most expeditious way. The hierarchy of management - the concept of the line manager - was appropriate at a time of slow communications and the involvement of large teams of workers. With rapid communications and shifting social expectations, company structures have become fluid if not fashionable. There is matrix management, "logical service units," and flat structures. More and more often companies are abandoning historical precedents. Typical of such companies is Oticon, a Danish company which manufactures hearing aids or, as it insists, "psycho-acoustic devices". Oticon takes its product as *the perceived sound* as opposed to electronic amplifiers: a rigorous application of customer focus. In Oticon there are no titles; people are there to do jobs - there is only one category of staff: responsible adults. Staff salaries are set by peer review. Presumably, there are still the standard functions to be carried out but it is not fixed who might be doing them on any one day. Oticon staff have no fixed abodes - no assigned desks. The place functions because of portable telephones and - at the last resort - e-mail. Oticon is not an isolated phenomenon. Marketing companies, architecture practices, computer service companies, all types of consulting companies are now disembodied organisations.

All this results from the way systemisation has eliminated much of the management of operations. The routine of making sure that things were running well, the management dramas of solving operational problems such as bottlenecks and the ebb and flow of supplies have been replaced by systems monitoring. Systems are implemented precisely to banish bottlenecks and ensure the flow of supplies ebbs when stocks are abundant and surges on automated request. Managers have moved up a conceptual notch: they are concerned with change. Management has become involved in projects which change things to a smaller or larger degree. This has its own special vocabulary of goals, ownership, schedules, milestones and so on. It is also more visible and therefore more tense. The expectations of flexibility and imaginative problem solving are higher as there should be no routine projects. "Routine projects" is almost a contradiction in terms. They become like operations. They are subject to detailed checklists and methods of attack which render them down to systematic approaches to be given to junior managers. The end result is that managers depend on change for continued employment. And so careers require a culture of ceaseless change.

Irony

Management is like sex; every generation believes it has made the definitive discoveries. The selling of information technology has led to the devaluing of previous accomplishments. The paper-based corporations produced the wealth which was available for squandering in world wars and arms races and which produced the recoveries of the nineteen twenties, forties and fifties. It is easy to be mesmerised by the overwhelming victory in the 1991 Gulf War and forget the scale of organisation in the Second World War. For example in the second week of April 1945, the allies flew 6,200 sorties, landed more than 15,000 tons of supplies, mostly fuel on forward fields. That month approximately 40,000 casualties were flown out of combat zones. The US 3rd Army alone evacuated by air 135,000 men in the last month of the war. Not a computer in sight.

It is easy to forget that the strategy of the 1991 Gulf War incorporated the lessons General Powell learned in Vietnam. The application of computers facilitated the tactics but did not define the strategy. The key data gathering device was airborne radar not ground borne computers. Cruise missiles, for all their spectacular demonstration of guidance, did not defeat the army in the desert.

Have computers transformed industry or are they just part of industrial evolution and social transformation? Are computers the agent of transformation or just the right bit of the jigsaw? What would have happened if Babbage's machine had been a roaring success in the nineteenth century? After all, nineteenth century industry could just as well have used them. The whole process of systemisation of business would have been accelerated with Babbage's machines. Systemisation is the under-studied aspect of the industrial revolution. As soon as automation was possible, a sense of system was required. From the 11th century, with its water and wind mills, through to the invention of the clock in the middle of the 14th century, to the automata of the eighteenth century, there was a growing sense of system and of engineering. The Jacquard Loom, and the use of steam to transport energy around a factory meant that devices became parts of a greater system of manufacturing. This is the key to understanding the proliferation of computers, in their role as the archetypal systemisation tool. To see their future is to understand the relentless systemisation of society.

The shape of modern society, with its defining characteristics of rapid transport, abundance of goods, easy access to education, an expectation of entertainment and mass communications and an elaborate social infrastructure, was set in the middle of the twentieth century. The computer revolution has added a thin top layer. Joseph Weizenbaum, a pioneer of artificial intelligence - or what he called "look up techniques" - wrote extensively on the limits of computing. Not only its theoretical limits but also social limits. He observed that "the computer was not a prerequisite for modern society". That it was only by its enthusiastic reception by well financed groups in the United States that it became a necessary part of the infrastructure of the military, the government and industry. In doing so it ensured an infrastructure that was dependent on itself. The political philosopher Langdon Winner observed in an essay "Mythinformation in the High-tech Era", written in 1989, that there was almost a total silence about the end or purpose of the computer revolution. To him the purveyors of the computer revolution seem "filled with the conviction… that society will renovate its structure to accommodate computerised, automated systems in every area of concern. The efficient management of information is revealed as the *telos* or purpose of modern society, its greatest mission."

The postmodernist philosopher Michel Foucault claims that every age is embedded in an *episteme* or standard mode of narrative and description. This, allegedly, determines the way in which we understand and describe the world. Foucault claims that history, the explanation of things as the result of an historical process - is the modern *episteme*. The "postmodernist" *episteme* is something else. Typical of postmodernism, the arguments are arbitrary; the explanation of social conditions by history is not typically "modern". An argument of some force could equally make the claim that it is systemisation that is the modernist or even the postmodernist episteme. It is not over-stretching the case to associate "progress" with the increase in systems. From the Middle Ages on, behind almost every material manifestation of progress there have been systems. Water supply and sewage disposal are necessary for liveable cities. A legal system that works is a prerequisite to control arbitrary power, and civil progress has been a story of curbing power. If we look at the social "landscape", the number of systems - social and mechanical - increases with prosperity and order. Yes interruptions occur, and despots pervert intellectual accomplishments to their own ends, but the trend of covering this social landscape with systems increases apace. Now we live on systems or wedged in between them. It seems there is very little social territory not fenced in by systems.

Why this impulse to systematise? The obvious reason is that it is strikingly effective in all material aspects of life. It has transformed not only the production of goods but, more generally, the organising of people to obtain a goal.

Effectiveness is a seductive criterion for the value of many activities, but effectiveness can only be judged in relation to a goal. The effect of systemisation on culture is to redefine problems and solutions. We often hear about a solution to a problem being "a technical fix", frequently a successful fix. Humans are no longer agents of intelligent free will, they are components of a system. A lack of goodwill, discipline, training or whatever, is taken to be a state which we simply have to accept, and redesign the system accordingly. An example of this is the education system. It is constantly redesigned to produce acceptable educational outcomes. The assertion that modern living might have lessened the resolve of children to strive is irrelevant. A system will take the "people component" as it finds it. The solution to the problem is to alter the system. The same is true of politics. It is realistic to suit the system to the customers, but if the system is itself partly responsible for the poor shape of the customers, the readjustment of the system might not be solving the problem we want to solve.

Chapter 2: The System's Embrace: Living by the Rules

When you move to a new company there are two challenges to overcome. The first is to create or re-establish a reputation for competence and diligence. The second, which can take longer, is to figure out the new employer's system: the forms, the rules, the correct phrases, the manuals to read, the "system police" to appease and who will kindly answer stupid questions. Welcome to the system.

The Socio-Technological Machine

The ultimate aim of systemisation is the organisation as a socio-technical machine. This repellent phrase actually describes the most productive and successful businesses. Large manufacturers, banks, supermarkets, large variety stores, fast food franchises are all of this type. They are arranged so that the customer can act as though the business was a giant vending machine. In the case of the variety store and supermarket, the customer is in an inventory system and the checkout counter is where "loaders" load data into the computer and debit the customer's bank. As soon as this can be done reliably by the customer, the checkout staff will be dumped. Banks, which are only as good as their computer systems and their outlets, including the distribution of their ATMs and EFTPOS links, encourage remote banking. Their staff are there to catch new business and deal with the tiny fraction of business outside the system. In manufacturing it is easy to have an automated system which accumulates orders, inspects them for appropriate "batching", checks existing production schedules and updates them. The system makes sure that the required materials are available, alerts despatch agents when the goods will be ready for shipping, bills the customers, and pays the suppliers and workers.

The more systematised an organisation becomes, the more it resembles a well-oiled machine, and the less there is a need for management. As automatic devices replace people in factories, those who manage the factory are replaced by operations "observers" who understand the devices which make up the factory. They are there to make sure things go according to specification; to ensure that functions function as only functions should. The factory system might even be so "intelligent" as to be contrite: monitoring itself and reporting some of its faults. Interactions between processes, and quality control are quite easily accomplished. There are sophisticated statistical techniques which allow machines to record and monitor their own history and, with a little human help, or plenty of bad runs from which to learn, arrive at algorithms to detect the onset of adverse conditions and take correcting actions. The organisation becomes a "megamachine " (to use a term coined by Lewis Mumford).

Crewing the device - learning the rules

The image of a modern organisation as a device - a socio-technological machine - seems at odds with the march towards offices with no fixed desks or rooms, with "flattened" corporate hierarchies and most workers being "knowledge" workers or in the service industry. But as far as organisations go, an employee is there for the skills that have yet to be automated. If it were possible to know the rate of systemisation of a skill then the possessor of the skill could calculate the time before they became obsolete. So far no one possesses this skill. The achievements mentioned at the start of Chapter 1 - the defeating of the world chess champion by a computer and the summarising of legal journal articles - suggest that the automation of high-level skills is not going to stop.

In the long run, we are all statistics. To get paid, to be an economic statistic, we must be seen to be doing something, and something for which someone is willing to pay. In an organisation this becomes very specific. To be paid I must be part of the operation of the system, or a "maintainer" of the system. I must respond and produce. As far as the system of the organisation goes, I must respond to orders, queries, schedules, even demands for useful ideas and I must produce something in response. I am a something that responds to messages with some action or I am nothing. Furthermore I have to do this in a way that a machine cannot replicate, cannot do, or cannot do as cheaply, if I am to remain useful to an organisation.

A system is about rules. There are simple rules, there are elaborate techniques to perform, there are methods, protocols and, if you need to sound like a consultant, "methodologies". In corporations, organisations, and franchises the world over, people carry out these rules, methods, protocols, techniques and methodologies. To avoid repeating the list every time, the word "rule" will be used. Rules are the heart of a system. They are marvellous things for ensuring that people are servants of the system and can be blamed for deviating from the rules. The creation of the rules is the subject of a later chapter but for now we shall take it that they are there to be obeyed. A detached observer - a child or an alien anthropologist - observing the nature of modern business might describe business as people responding to messages. A message is just a general way of describing something that prompts someone into activity. It can be an order at a fast-food franchise, a request to design a building, check an insurance claim, paint a house. A message produces actions which usually result in other messages. Something is to be done, something is needed to complete a request, invoice an order, approve a plan, enrol a student, treat a patient, prepare a budget. The more the response to a message is constrained by rules, the less intellectual work is required to complete an action.

The table below shows some examples of common types of rules.

Type of rule	Examples
Definitions and rules for recording data. What is to be recorded and in what way. These rules are enforced by the design of forms or computer screens. They are apparent when computer screens beep and refuse to let you carry on until the correct type of data is entered into a field.	To be a student a person must be enrolled in a course which, in turn, is part of a qualification. Your PIN number must have at least five digits.
Operational Rules: These are requirements which must be satisfied so that operations comply with the desired standard of practice, including safety.	Every course must be assigned a time, place and lecturer before any enrolments are accepted. Flammable or explosive cargo must not be carried on passenger flights. On the night shift there must be at least one registered nurse for every 10 patients. For all master files there must exist two backup copies, no older than 24 hours.
Rules for dealing with the allocation of resources to match demand.	These rules are involved in the creation of timetables and the scheduling of extra staff, trains, dynamos, policemen etc. in peak times.

Rules of notification ("triggers" and message generators). Under what circumstances people should be notified.	The number of people in the waiting room should not exceed 10. Whenever it does, the receptionist must notify the supervisor to switch staff from other duties. Operations staff must be told of the next schedule whenever there are less than 5 orders left to process.
Rules for change of status or change of data in a record. Rules of notification can be linked to this.	A supplier with more than three missed deadlines is re-ranked as a "casual" supplier. Change in a person's contract on promotion.
Recipes and Protocols. These are sequences of steps which lead to the completion of some task.	Factory operations, Checklists, "Methodologies": the approved set of steps to follow to achieve something such as the design of a specific product for a customer.

The rules of a company grow by accretion with a passing parade of managers trying to bring order and better practices to their operations. The accreted pile can be formidable. In 1992, at the time Arthur Martinez became head of the corporation, Sears had a rule book of 27,000 pages. Martinez replaced it with a 17 page code of conduct.

Any system which relates individual things, such as employees, pay-rates, superannuation payments, orders, products, inventory items, patient's drug orders and so on can be expressed in terms of *first-order logic* . This is the logic which relates things satisfying one property with things satisfying another. The relating part, the relation or relationship, can be thought of as a cross-reference table. The left-hand column is a list of individual things satisfying one property and the right-hand side contains the corresponding things satisfying the second property. For example the left-hand side can be a list of parents' names and the right-hand side the corresponding children. One parent can be listed with many children. Relationships can link parent with children, people with their addresses, customers with their orders and a product with its components. A relationship which uniquely associates items with one property with an item with a second property is a function. This occurs if children are associated with their mother. Each child is associated with one and only one mother. A relation which associates a mother with numerous children is not a function because, given a mother's name, you cannot cross-reference to just one child unless you add another criterion such as the youngest child.

Each time you use an ATM, your PIN and other identifying data on the card are related to your set of transactions as a relation, but to the most recent one as a function. These types of relations and functions are the stock and trade of relational databases, the whole theory of which can be written in first-order logic. It is called first-order because there is no hierarchy of properties or relations. First-order logic does not allow properties of properties or relation of properties. All properties are of things at one level of abstraction - concrete things if you will. Relations are between sets of concrete things. We cannot refer to arbitrary sets of things and certainly we can't say for all properties with a given property. That is the domain of second-order logic or *many-sorted* logics.

Relational databases are important in our story not only because they are ubiquitous in organisations which use more than word-processing, but because of what they demonstrate. They have demonstrated that an enormous array of business and government activities can be described in their terms. This means that the administrative logic of most organisations can be described as a vast system of relatively simple relations, all able to be written in first-order logic.

First-order logic is useful because of its restrictions. If we want to claim that a relation exists between sets of things which have some of many different properties, we have much more work to do checking properties for sets of things than for individual things. If we want to check a property of properties as opposed to a property of an individual (concrete) thing, then the chances are there is not going to be an easy way of doing it. For example it is easy to check whether a person is an employee, or a barrel of wine has an alcohol content greater than 18 percent, than it is to check that the property of being over 50 years of age is "undesirable" - that is, if it is true then the *property* of being over 50 years of age is undesirable. This is not a statement about any individual of any age but a property of a property. Similarly, to say that a relation is a strictly increasing function, which would be a desired property of a sales graphs, is a desirable *property,* is quite different to being given a strictly increasing function. The second or higher-order logics deal with sets of properties as well as all the things with which first-order logic deals. A set of properties can of course be a short list which can be dealt with in first-order logic. It is when the set of properties is not given by a list but by a property of properties that we go beyond the scope of first-order logic. The power of these logics is considerable. It is like an advance from steam to electricity and in a constrained way they will be a part of the future of systemisation.

Systemisation is the enforcement of logical systems in a social context. By embracing production and administration systems, society in general - and business in particular - is doing little more than exploring the manifold applications of these logics. This exploration has no clear boundaries. It seems that ingenious ways are found time and again of rendering into logical form, tasks which seemed secure from automation. Not only do systems get more elaborate, but the social activities are moulded into ways which can be rendered into systems. This is seen by the way innumerable transactions, such as paying bills, buying tickets, can be done without human interaction. It is also seen in the change of the descriptions of services. Training is now seen as producing specific behaviours which are called outcomes. The administration of hospitals is also infected with "case equivalents" (medical and surgical procedures of a given type) and outcomes.

Algorithms and Experts

Systems given by rules are not necessarily comprehensive. The rule "use only high-tensile steel in this product" does not give any idea of whether the high-tensile steel is on hand or not and what to do if it is not. Systemisation works towards a comprehensive description of procedures through algorithms . An algorithm is a recipe for doing something including what to do in the case certain things are not available or not in a required range. To be automated, rules have to be able to be cast as algorithms, which requires step by step evaluation of states. The state of a device at a given time is the setting it has at that time. Inputs and the recent history of the device must determine the states unambiguously. Once a state is checked, the next step has to be unambiguous. The discovery of algorithms is no trivial feat. Cruise missiles are controlled by complex algorithms. Computer games are complex algorithms. Deep Blue, the 1995 "world champion" chess program, is an algorithm which contains an algorithm for determining which is the best of millions of options. The difference between algorithms and functions is that a function can be completely described without describing how to calculate it. An algorithm *always* calculates a function. It is a completely detailed recipe for doing that calculation. It is the difference between a rule requiring a skill and a "no brainer " - something quickly checked and requiring no interpretative skill. The McDonalds hamburger restaurant chain is founded on this premise.

The intention of most corporate executives is to achieve the greatest possible throughput. In the context of information, this would translate to minimising the flow of messages for a given function. Ideally a function would be automated, so the flow of messages is hidden inside a "black box". This means that the function can be incorporated in an algorithm. Among all the possible algorithms there is at least one which minimises the flow of messages. This would then be the quickest message-processor for a given function. However, most functions have a residue of uncertainty. Dealing with uncertainty is, in some ways, the core of expertise. Conversely, no uncertainty implies that problems can be dealt with by rules; by algorithms. For example, if I know the state of a customer's mind, I do not have to guess the upper limit of price he or she would pay. If I knew where the customer was in the cycle of checking the options, or what they were really complaining about, then solving their problem, to my advantage, would be simple. We make all kinds of judgements as to what people really want, what would really comfort them or satisfy them. We make guesses as to the state of traffic, the number of customers we might receive, whether a machine will last the next 1,000 operations. By adopting margins of safety, by building in excess capacity, by adopting regimes of maintenance, of training, and of corporate behaviour, and by following endless methodologies, managers seek to remove uncertainty and risk, and to increase the domain of application of rules.

The pursuit of throughput at minimum cost has led to systems which minimise judgement, hunches and guesswork; systems which apply algorithms as much as possible, systems which automate as much as possible. Though this is well beyond the dreams of Frederick Taylor, it is the history of business in the twentieth century. It is the reason for the huge gains in productivity that the century witnessed. It is also the reason that books on management now assume that organisations are filled with "knowledge" workers, staring at computer screens, and making decisions of varying importance.

Automation requires that messages or problems are classified into sets to which the system's rules can be applied. Once a message is classified, which is an interpretative task and might require someone as skilled as a lawyer, it can be submitted to a rule. This is a typical application of artificial intelligence techniques. How does this alter the organisation? The arena in which expertise is used is originally in classifying problems. By reducing this to a minimum, expertise is isolated to non-algorithmic "judgements". In routine operations there are very few cases where a check sheet or method is used and where there is also a requirement for judgement. Check sheets and routine methods are designed to eliminate judgements. Check sheets, for example, are frequently used to classify something into the domain of some rule. Classification is a necessary act when setting up a system. It defines when an individual satisfies a property such as being a student, a commercial customer, a manager, a contractor, under which conditions oil is too thick, too dirty, too inflammable and so does not have the property of being re-usable. Once this is done, a method can be applied to solving the problem or recommending a service.

As an example, consider the problem of whether to invest in property in a particular area. This would look like a difficult decision. However, the answer depends on a series of simple questions:

1. Can we afford it? (Yes if it costs less than $450,000)

2. Will it benefit us?
   Is it something which will gain in value?
   Is the area going to be important in the future?
   Is it likely to be an area which attracts further investment?
   Who is already investing?

Are they big?
What is the present rate of investment in the area?

3. Will it benefit our suppliers (or friends)?
     Is it easily accessible?
          Does it afford good customer parking?

4. Will it produce more customers (or friends and relatives)?
          Will it make attracting public attention easier?
               Will it make us look more professional?
5. Is it cheaper than other options?
     ......

We can imagine that the weighting decreases from 5 to 1 according to the depth of the indent. The expertise - such as it is - is in the construction of the questions. For old hands at investment decisions, such questions are second nature. They can be thought through in seconds. If all the information is available in some database, a computer can also decide investments in a fraction of a second. This is what already happens in international commodity and foreign exchange markets. Much of the expertise of the commodity or foreign exchange trader has been devalued to the cost of a small computer program, a case of systemisation leading to devaluation. When this happens to a skill, it ends up in an educational bin marked "taught but never used". What is true is that it is taught because it needs to be understood by someone even if it has been automated. Just because producing matches is fully automated does not mean that no one has to understand the process.

This example is simple but the principle behind it is far reaching. Many highly rated skills can be split into sequences of simple questions. The classification of a message, or a problem, is accomplished by answering the questions. This produces a sequence of yeses and noes which can be used to code the class

51

of the message or problem. There might be a thousand questions, but given sufficient motivation, if they can be discovered, an automated system can be made. Effort, insecurity and ignorance of the possibilities combine to inhibit this being done more often. Again the automation of a process or skill does not mean no one has to understand it. The important knowledge is transferred to the understanding of the automating.

Skills beyond the System

Humans are required whenever the complete enforcement of rules cannot be accomplished or when logical description falls short of what is needed to describe an algorithm. These cases are the source of non-computer decision making, including the classification of individuals or events within a system, and decisions to create, review or change systems. Various levels of management are called on to do these tasks.

If something can be easily taught, such as the operations of a fast food outlet or changing the oil in a car, then it is a short sequence of "when this, do that" instructions. The employee is not expected to make judgements which require any length of experience. If they do make useful judgements they can be promoted to supervisor. Then they will be required to make sure that the overall operation of a group of people is satisfactory: that the set of actions completed by the operators are within expectations. Further promotion to manager requires the co-ordination of sets of different activities which all must come together for a part of the business to be successful.

Charlie Chaplin's "Modern Times" (1936) showed the worker as a cog in the machine. Managers felt secure against being replaced by a device until computer systems made a layer of them redundant in the 1980s and 1990s. In terms of systems, jobs can be classified in terms of the highest level of logic they use. If a job can be fully described by first-order logic, then it is under threat of being done by a machine. Status accrues to jobs the further they are from having a complete description in first-order logic. The refuge of human judgement is the area which computers cannot conquer.

But the puzzle, the nagging insecurity, is how much of a skill is really outside the domain of algorithms and computers? How much of that skill is beyond precise description? The first part of the process of creating an algorithm is the precise description of all the ingredients of something to be computed. Computing the cost of a product is easy if every component, including labour, can be listed and costed. The rest is simple arithmetic. Artificial Intelligence is an area of programming where new ways of describing activities and phenomena are invented for the purpose of simulating them on a computer. These descriptions can be as esoteric as descriptions of sound-waves which occur in birdsongs. Once a sufficiently comprehensive set of these is available, they can be represented on a computer and an automatic identification of birds from their songs is possible. Computer speech is advancing with a similar idea of compiling huge numbers of sound patterns and then linking them with another formidable list of responses. It is not a description of the way humans talk, but it exploits the abundant power of the computer to look things up, relate them and string together a response. Summarising articles is similar. In the simplest sketch it is done as follows. First of all a "semantic map " is created. This is list of nouns, adjectives and relations between them. Sequences of these are generally ranked for importance. The more extensive this map the more it captures a fragment of expertise. The summarising program then "reads" the article, checking for the high ranked nouns or phrases and lists the sentences where these occur. It also needs to be able to figure out which sentences are roughly similar so it doesn't produce a large number of repeated sentences. Recognition can be accomplished if the objects in the area to be considered can be described in a way that the computer can match the input with a stored description of the object. Cruise missiles testify to the extent this can be done. But accurate descriptions of the terrain need to be available.

As a rough guide, those areas safe from systemisation seem to be in recognising things, intentions and purposes, in issues of motivation, emotions and conceptual creativity.

Algorithms to recognise faces and nuances of expression are poor. The reason being that we do not know how we do it. We do not know how we make conversation or jokes. Witty repartee, conversation to put someone at ease, to soothe or comfort a person are highly creative, everyday acts. They involve the whole body in ways that we dimly discern. The description of these social interactions is not going to be complete. Whenever someone describes common sequences of movements or words in these interactions, and this becomes common knowledge, those sequences are avoided less a hint of insincerity is given. This is common in many human interactions from selling to the mating game where parties can be alert to insincerity and clichés.

Of Management and Expertise

From the point of view of systems, expertise is the ability to go from a problem, or starting state, to a solution, or desired state or result, with the minimum of time and effort. An expert portrait painter goes from a blank canvas or paper to produce a recognisable portrait with only a small number of mistakes and re-workings. (Well it used to be like that until art spent a century divesting itself of recognisable expertise.) An expert cook, systems designer, manager or architect navigates rapidly through distractions, irrelevancies, wrong and poor choices to a result which satisfies. Making mistakes, hesitating, confusion and making poor or wrong choices is the misery of a novice and it is the elimination of this which leads to high skill or expertise. In an organisation, expertise means capability and efficiency.

Expertise determines the capabilities of any organisation. There is expertise to use various types of machinery or technology. The term "technology", which seems to have been captured by the electronics, computer and telecommunications industries, includes drug regimes, pesticide spraying regimes, manufacturing and agricultural technology, mining equipment and so on. Wherever some device is used to make things happen then someone has to install it, maintain it and make it work. That is technology or device expertise. The level of technology is the level of making things work. It is what produces the goods. Whether it is a programmer coding a set of specifications, a motor mechanic servicing a car, a nurse inserting a catheter and setting up a drug pump or a coffee shop attendant making a cup of cappuccino, the person must know how to work the technology, for it is at this level that the money is made. The customer wants the fruits of the technology.

A person who can use a device well might not have much understanding why the device is constructed the way it is, although this does not hinder them from using it well. This would be true of many devices found in homes and offices. It is less so of those found in factories. Expert device users usually have some understanding of why it should be used. For example air-tools in an automobile workshop. The person who can figure out what device or technology should be used to solve a problem is a *technique* expert. People who earn their living putting together useful devices to solve some problem have expertise in some of the techniques of a field. The designer who can quote data for using piston or turbo-prop engines in an aircraft, rather than jet engines, is at the level of technique. That level of expertise might not be very good at dealing with piston or jet engines but they understand which technology will solve what problem. The level between technology and technique is the supervision of the technology - a person who can check that the operations of technology are satisfactory and, if not, decide what could be done to improve them. The person who has knowledge of technique is useful only to the extent something is to be changed. They have a creative role in applying techniques to solve problems. They are in the position of managing expertise - both co-ordinating and appraising.

A higher level of expertise is the level of principles. A person with expertise in principles understands why one technique is to be preferred over another. They can, for example, explain the aerodynamics of piston engine efficiencies as against jet engine efficiencies. In computing, the ability to implement a system design in a given database product is a level of technology. The level of technique is the level of database design. The designer might not know much about the technology in which the design is to be implemented but that is not their level of expertise. At the level of principles, the best computer representation of the problem is the concern, a database design being just one of the options. The person who works at the level of principle is the policy maker. The techniques represent what is at present possible and the long term goal is to achieve beyond them. The person who has expertise in principles should be good at the technique level but comes with no guarantee at the level of technology; biochemists and food technologists are not necessarily good cooks.

The following diagram illustrates the relationships between the components of expertise.

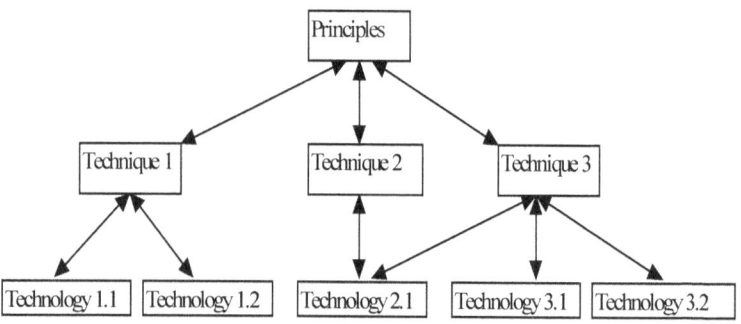

The double headed arrows are there to emphasise that communications must be in both directions. That good understanding of technologies grows with the understanding of techniques which itself enriches and fleshes out the principles. Nothing illustrates this better than business itself. For example, ideas such as payment via electronic funds transfer at the point of sale (EFTPOS) and automatic teller machines, which take banking out of banks, develop with new techniques of distributing bank transactions. This widens the very idea of where banking business can occur and how to reach customers. In high technology manufacturing such as aircraft, the pilots - the technology experts - provide the data which the engineers and physicists use to refine designs and even elaborate the principles or aerodynamics. The invention of new versions of an existing technology, say a new racing motorbike, proceeds from an idea in the designer's mind, to specification of components, through layers of detail to the actual bike. The rider is the technology expert not the designer. Improvements usually proceed from thinking at a level including both principles and techniques. The ability to innovate at a fundamental level starts with an extension of knowledge at the principles level. Much of the huge range of modern materials has come from chemistry laboratories investigating polymers and various types of fibre-based materials. In most cases the application of the research is clear - the extension of techniques is expected to follow automatically. This is an example of techniques being targeted for improvement. It can only be done with people who have a higher level of expertise: the knowledge of principles and their relation to techniques. The number of people in the top tiers of expertise, the technically and scientifically educated, ultimately determine the wealth of nations. Entrepreneurial flair, the social tolerance of taking risks, the willingness to change and dispense with obsolescent techniques, are the commercial side of the innovation coin.

People skilled in techniques tend to move towards principles as their jobs are automated in special case after special case. Many accountants have scrambled into financial systems as their cost accounting jobs have been eradicated by computers. Factory workers are either promoted into supervisory jobs or made redundant when technology eliminates their jobs. Relentlessly, devices become a sequence of components, each component being close to the optimum for the relevant set of principles. Construction, whether of photocopiers, buildings, aircraft or computer systems, incorporates larger and larger components, produced with increasingly automated methods. Every time the capability of components increases, more expertise is embodied into machines and processes. Thus systemisation causes the depreciation of expertise. The more the use of a technology is systematised, the more the expertise in that technology is reduced so that the operator is a "loader", a supplementary device interposed between the materials or between the customers and a device. As mentioned before, status and pay decline as the expertise is reduced.

Where this three tiered model of expertise fails, employment increases. In tourism, social services, entertainment and some marketing, principles, technique and technology or practice cannot be separated. Employment rises in these refuges from systemisation.

Techniques and technologies depreciate with time as does the expertise associated with them. This use of techniques and technology is as much open to influence or fashion as art or entertainment. As the company associated with a technology prospers or stumbles, the value of expertise in its technologies rises or falls. In a world of abundant "technological solutions," investing time in learning a technology is tantamount to investing in a company or group of companies. A parent or teacher advising a child to adopt a particular course of training is in the position of an option trader buying a portfolio of technology futures. Futhermore, trends imply that except for the most qualified, or people in very senior positions, career prospects over fifty become very constrained. Don't expect promotion after that. Long term careers are vanishing. Companies pick up - invest in - young trained staff, and whenever possible avoid training staff, except when they can "mould" them as well. Except for spot shortages - often the result of insistence on a very narrow specialised set of skills, experience, age and cultural background along with the no training and certainly no retraining policy - skills are in abundance. Universities and technical institutes are driven by the wishes of their customers. The result is that graduates are produced in excess of demand and this makes things easy for companies.

This is a short-sighted tactic. A company which fails to innovate and train is in danger of becoming a "device" itself. Something which just goes through the motions. The capabilities of a company depend on creativity built on expertise. Expertise planning and development, the link between planning and human resources, usually assumes that the bank of expertise is the community. The expertise withdrawal charges being the cost of luring someone from the opposition. This is an unusual philosophy for many a company which plans its material or capital requirements minutely. Expertise is *the* business commodity which determines what can be done - what the organisation can actually accomplish.

Managing the Device

The hierarchy of skills can be applied to the skill of management. What do managers do - what is the skill of management? The role of managers of various levels should be well worked out. Books on management frequently concentrate on motivation and leadership. But what is the role of a manager within the system? What do they do that cannot be done by a robot?

At the operational level managers are there to allocate work, control quality and enforce what might be called organisational culture. The first two can frequently be done by a machine. The allocation of work can be accomplished by computer programs.

Ascertaining quality can be difficult when the work to be evaluated is a report, a project, or a service. Some clear measure between what is good and what is bad is required for this to be automated. Most work that involves more than a standard output has nuances which elude simple measures. Where there are simple measures, managers are obsolete. The enforcement of organisational values is really making sure that organisational rules have been kept. These include good interactions with customers, clients and suppliers, punctuality, keeping deadlines and so forth. It is the management of people. This is the stuff of operations management: if something is supposed to have some level of quality then the manager's job is to make sure that it does indeed have that quality and to correct things when it does not. On the scale of expertise, this is knowing the technology. The technology is the way the company, thought of as a machine, creates its output. It is the operational system of the organisation. The operations manager supervises this technology. In a military situation this person is in the field, maintaining discipline and co-ordinating operations. They have their battle orders but did not choose the battle.

Above the level of "technology operations" is the level of technique. In this case the technique involves the choice of what "technology" the company is going to "be" in order to accomplish its goals. This level of management assesses the systems which are going to be used to obtain a goal. The military analogy is mapping out a campaign: where to fight and with what. This is the strategic level. The broad goals are worked out. The strategic manager has to figure out how they are going to be accomplished. In day to day terms, this is the constant reviewing and assessing whether some technique or some system is a good choice to achieve goals.

Management expertise at the level of principles is the top level of management. The general principles of the business are known. The broad issues of the relevance and choices of technique need to be understood. The details of the specific technology - the day to day operations - need not be. Although it is obvious that the better the understanding of the technique, the better the understanding of the technology. (This suggests that management skills that are only "generic" are only the start of a good manager's skills). This level of management combines formulating policies and thinking through strategic options. It decides restraints and rules out some strategic options. In the military examples, it is decisions to use coalitions and what weapons will and will not be used. It should outline the sequences of actions which the strategic managers must plan in detail.

Just as with any expertise, the lower levels of management are the most vulnerable to being systematised out of existence. The 1980s fashion for downsizing eliminated layers of middle management as it was realised they were little more than a gathering place for data. This started when systems which produced large reports which then had to be summarised, changed to systems which allowed users to define many of their own reports. Jobs which required little more than pouring over pages of printouts looking for things to query were no longer required.

The description of management in terms of functions and levels of expertise does not depend on any particular corporate structure. One person might do it all. The description depends on the idea that the organisation is a device or a technique for accomplishing something. The size of the organisation might dictate many departments and specialities, and hence many managers, but in terms of the systems there is little more. This is so even for human resources - by the very choice of its name, it slots itself into the system. They are there to get the best use of the bits the automated parts don't reach. That these bits are skills that come packaged in humans only means that the concentration and focus of the humans has to be kept up. So the part that human resource departments manage in the scheme of things is the supply of concentrated application of skills, both humble and mighty. That this seems overly cynical is only to emphasise the place of skill in the system of the organisation. That clever personnel, human resources, human capital or even "talent management" staff contribute to convivial, humane corporate environments should not obscure the fact that they have failed their system role should the supply of useable skills become inadequate.

This picture of the manager technician seems removed from the familiar picture of managers stirring the corporate pot, restructuring, reorganising and re-engineering projects, staff, furniture and paper trails. The reasons for constant change are examined in a later chapter

Computer Illusions in an Age of Systems

Surely this bleak description of the organisation as a system is inconsistent with the modern office with its creativity, fun and computer on every desk. These computers are there as servants to help the work flow and allow us to communicate with our colleagues. But would you get a computer on your desk if it did not make you more effective as a unit in the system? And for a long time - to most people's surprise - the answer is "No". You did get a computer on your desk for no clear reason. But first a little history.

Early computers were very expensive and data storage was largely limited to magnetic tape and magnetic drum storage. Tape, in particular, forced files to be read from start to finish. This meant that applications were largely mathematical or long sequences of calculations. Payrolls, social security payments and accumulative accounting functions were the early administrative applications. While this produced some increase in office productivity, it was the application of operations research that drove much of the effort in the 1950s. Operations research was developed in the Second World War - at least to the point that it had been named and described. It had contributed to such diverse areas as the deployment of radar, strategic bombing, aircraft maintenance, submarine warfare, code breaking and the nuclear bomb project. In particular, during the 1950s and 1960s, the possibility of nuclear war accelerated the development of Game Theory (see Chapter 4). Operations research also spawned the new subject of Systems Analysis. Systems Analysis was initially associated with the Rand Corporation, a highly influential private think tank in the 1950s. Rand, standing for Research and Development, employed people such as Hermann Khan, variously described as mathematician, strategist and futurologist, and Albert Wohlstetter (less flamboyantly just an economist and mathematician). Khan's particular claim to fame was a series of studies on the effects of nuclear war. Although this earned a great deal of odium, it had the virtue of initiating a serious study of what nuclear war could mean. Wohlstetter was commissioned to work out where the United States airforce could most economically locate its overseas bases. His report, originally taken as an accounting exercise, looked at such factors as the economics of enemy damage and so placed the report in a fully operational context. Systems Analysis is probably the most influential, abstract technique ever to be embraced by practical businessmen and women. It is also unsung and unglamorous. Its many manifestations and guises underlie the reliability and safety of successful social, engineering and computer systems.

In the 1950s, large industries such as the petrochemical industry, airlines and manufacturing companies seeking to minimise waste in the use of time, skills and raw materials, applied and developed operations research techniques. This would have delighted F. W. Taylor as these techniques were a mathematical approach to optimising components of the organisation itself. The most highly publicised techniques that were developed at this time were linear and "dynamic" (essentially non-linear) programming associated with Kantorovich in Russia and Bellman in the United States. Kantorovich was considered crazy by his colleagues but as he won the Nobel Memorial Prize for Economics for his labours, he had the last laugh. Much of this work was commissioned by the armed forces and the space agencies. The flowering of operations research grew from the discovery that mathematics could be useful "in the real world". Its association with computers was fortuitous, a happy marriage of convenience which exploited calculating speed, but not a necessary arrangement. Indeed it is rare now to find much operations research knowledge in information technology specialists.

The modern phase of computer applications really starts with the widespread application of databases which only became effective after the invention of the disc-drive in 1956. This seminal invention, first produced by IBM and known as RAMAC for Random Access Method of Accounting and Control, thus including its target market in its name, quickly replaced magnetic drum storage. It enabled the development of relational databases after the theoretical work of Edward Codd of IBM in the early 1970s. These developments, and the development of smaller, cheaper but more powerful computers, led to the ability to access information "at random", something that was necessary for "online" applications such as automatic teller machines and systems which allow immediate access to customer/patient/client/student records. Unlike operations research, these ideas are the common currency of all information technology specialists. Databases meant that administration and marketing became the prime movers of organisational computer systems. Companies developing information technology knew that these powerful groups wanted access to records and played to them. Thus database suppliers and administrative computing became "IT" until marketing discovered the Internet in the 1990s.

Since the 1970s, manufacturing concerns, banks, insurance companies, supermarkets and hospitals have standardised computerised administrative systems. These companies still lead in the purchase and elaboration of large expensive systems. It is worth gaining some idea of the ingredients of these systems as they are a representation of the integration of businesses. They are what is driving the device aspect of corporations, making them more and more machine like.

A standard manufacturing system will have the following ingredients

(In this exercise in Systems Analysis the convention is that the boxes signify processes or actions and the arrows signify a flow of goods or information.)

In addition to this, there are the accounting, financial and asset systems, the sales and marketing systems with their incentive systems, payroll and personnel, and a host of other systems which support the main ingredients, such as design systems which design or configure the desired products and cost them automatically. Service industries are similar. An order is often drawn up with the aid of someone who can assess what is needed to provide the required service.

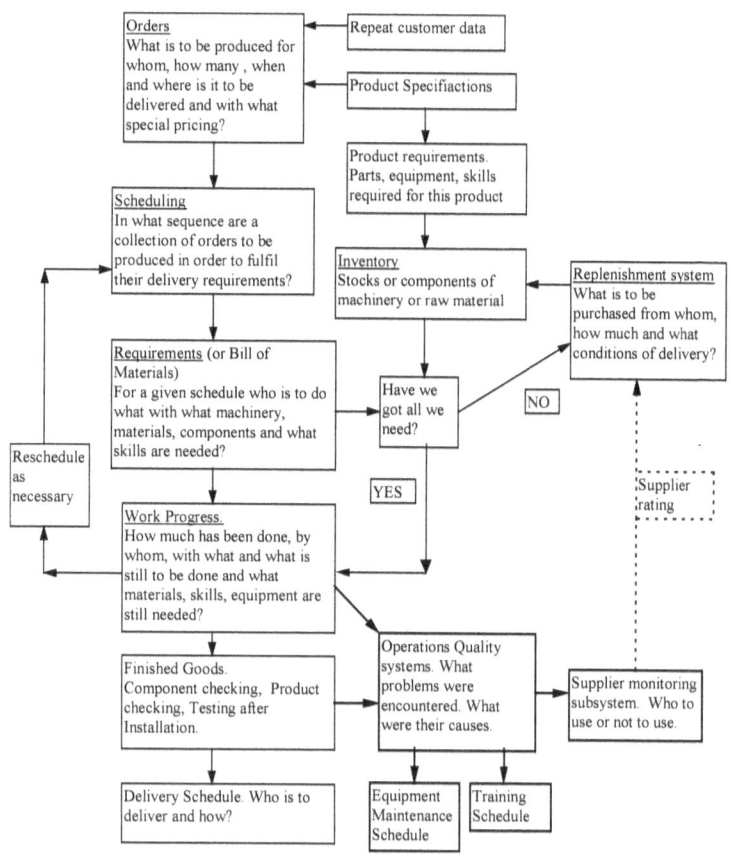

When the customer is in bad shape, such as when shipped to a hospital, a doctor works out the order: it is called a medical plan or treatment plan and details what procedures will be necessary to fix the patient. The hospital might not have the required skills or equipment on hand and so the patient has to be moved to a better equipped hospital (or take their chances). In banking, an order is simply the opening of an account and the expectation of professional care of the deposits. Other types of orders are borrowings.

These systems are administrative systems. They do not make production happen faster. They do not transport people, heal them or educate them faster. The might remove some of the form filling and chasing of information that delays the production or the service. They all require the same information which was in paper systems, to be entered into computers. They do make it easy to control costs and the activities of people. The "timely, accurate information" slogan which led to the development of these systems has produced a closer scrutiny of people's activities; how they use their time. It has allowed the examination of the performance of people and organisations which impinge on the business. And, in the case of the banks, it has allowed them to do previously undreamed of things, such as provide services almost everywhere, all the time. It has allowed airlines to provide world-wide services, a commodity market of tickets and frequent flier points.

A Boston Consulting Group study of large integrated software systems found that only a third of them had "created value", were cost effective or had tangible financial impact. It anticipated that the companies in the business of major information technology projects will have world-wide revenues of $US125 billion by 2001. It also found that 12 percent of companies had rid themselves of the software purveyors. One in eight does not sound like much but given the investment in money and reputations it takes a great deal of exasperation to change software horses in mid-stream.

Databases are the core of administrative computing. Stripped of glamour, databases are, as briefly mentioned before, just lists of related items. They replace paper lists as easily looked up references. How are they related to the rules and messages which lie at the core of systems? Each time an order is recorded, an invoice is raised, a patient recorded as admitted to a hospital, or an insurance claim entered into the system, a set of rules has been followed. A database structures the recording of events according to rules. The order has to have certain data recorded; an invoice must have a customer and an order recorded on it; a patient record must have a description of symptoms, the time and date and the admitting staff member; the insurance claim must have a policy reference and details of the incident in a prescribed form. Databases record the activities of the organisation in a way that the structure of the records reflects the rules by which the business operates. These records are the contents of the messages which are the daily diet of business. Databases split out the information in a way which makes it easy to search for the records of all messages satisfying a particular criterion and each search generates messages about messages - all orders for a particular item, all patients admitted last Wednesday with diarrhoea, all claims for fire damage in a given neighbourhood in a particular week. Managerial information can be drawn from this history. For example, if the growth rate of a section of a company is poor over a period of time, something easily detected from the accounting records, then a message highlighting that fact can be automatically sent to divisional management. Previously this might have been communicated by telephone whereupon the hapless manager might have had to account for the situation on the spot. The wonder of the billions of dollars of information technology, including databases, is to allow agents to look up lists of data in order to generate the messages they are paid to produce. It also allows the implementation of digital currency via EFTPOS. In other words, computer systems take care of all the intermediate, completely rule-bound parts of generating messages. It is considered strange that this has not shown up as a stunning improvement in productivity. It is a repeated refrain of managers that the money spent on integrated systems, which promise more and more, provided only the most marginal improvements.

The company ("enterprise") wide systems are the basic administrative glue of medium and large companies. Even more widespread are the Office Systems: wordprocessing, spreadsheets, presentation and graphing products and, of course, the Internet. These have been so well sold as to define a level of "professionalism". For the most part their overall effect has been to raise the standard of presentation. Presentation is now strongly linked with "professionalism". This has little to do with productive action and the quality of the actions and probably hinders them. But it is now so well entrenched that it is churlish to question it. Of course the ability to use the Internet is linked to an additional notch of professionalism. Again this has nothing to do with production or the quality of important services. It is simply taken to be "a good thing" on its own.

The Productivity Gap

The early expectation of computer generated administrative productivity, so loudly proclaimed by information technology vendors, did not show itself in the overall statistics of economies. The surge in white-collar productivity seems reluctant to announce itself on a wider scale. According to the United States Bureau of Statistics, between 1960 and 1973 non-farm output per hour grew at an average annual rate of 2.9 percent. From 1974 to 1975 it was 1.4 percent and from 1996 to 1999 it was 2.6 percent. The lack of increase in productivity on the wider scale was first noticed in the 1980s and has continued to be a curious dissonance in the digital elite. It was publicised in a number of places by Tom Forester, T. K Landauer, Landon Winner and in articles in Scientific American, November 1994. Forester noted in the early 1990s that, "Over $300 billion a year is now spent world wide on computers and communications hardware and software, but it is doubtful whether more than 300 researchers around the world are studying the impact of all this spending on the economy and society at large."

In 1990 Paul David wrote, "The early 1990s saw a new questioning of the value of computer systems - not just their cost to people in physical problems, job degradation and job loss, but also their economic return. Economists added up the investment in information technology - about $1 trillion over the course of the 1980s and about one third of all corporate capital spending -

and wondered what effect it had had. William Bowen, writing in *Fortune* in 1986, suggested that white collar productivity grew more slowly after the introduction of computers than before. Citing figures compiled by Stephen S. Roach, an economist at Morgan Stanley, he reported that white-collar output per worker hour had not changed between about 1970 and 1985, despite an enormous investment in computers in those years. Paul Strassman sums up the evidence in *The Information Payoff*: "The stagnant labour productivity numbers simply show that even in America's two most automated service industries (banking and insurance), there is no discernible effect on labour productivity as a result of extensive investments in computers." "We see computers everywhere but in the productivity statistics," commented the Nobel Prize-winning economist Robert Solow. Productivity was not going to make any sudden movements in the 1990s: "In the decade to 1992 service companies in the US were reckoned to have spent \$860 billion on information technology, while the service sector (which employed three quarters of the American workers) was only increasing its productivity by 0.5 percent a year - compared to 3.8 in manufacture". A survey in *Fortune* magazine in 1991 found "the most productive companies tended to spend less per employee on management information systems than did companies with average productivity."

William Bowen and others have suggested, as a way around what he characterised as the "puny pay-off from office computers", reducing the information system overhead - the necessity of dealing with complexity - by redesigning office systems, products and manufacturing processes so that skilled individuals can hold more of it in their heads. This is echoed in a 1989 MIT report that concluded, "We have spent over a decade and millions of dollars developing elegant Materials Requirements Planning Systems, while the Japanese were spending their time simplifying their factories to the point where materials can be managed manually with a handful of Kanban cards". This requires training and some commitment from management and workers. It is a well tried recipe for producing productive workers and staff who feel confident with dealing with customers. Such commitments and attitudes come and go in fashion. Complexity is only tamed when it is turned into sets of rules. A large set of rules undermines a person's confidence to know them all and forces the automation of the system. Workers are made redundant at the cost that customers are provided with a different type of service. The self service of petrol stations is such a case.

It is no surprise that office productivity has changed little. In most offices, especially those which have had mainframe or mini systems since the 1970s, the pattern of the 1980s was to produce reports on top of reports. Powerful members of the administrative staff requisitioned PCs or workstations for less and less critical reasons. An enormous amount of analysis was done; finer details of control were achieved but for not much payback. The office itself was not changed except for the illusion of better decision making. As no new information is added, very few decisions contain any new real insight. Furthermore, the nature of decision-making is not clear. Data are taken as the bedrock of decision-making and then arbitrary weighting factors are introduced in spreadsheets, to quantitatively justify prejudices. A number of studies have shown the extent to which spreadsheets contained errors. The studies gave error rates from 11 percent of spreadsheets (Computer Audit of HM Customs and Excise) to 30% percent in a study by R. Panko of the University of Hawaii, to a pessimistic 90 percent in a Coopers and Lybrand study. On top of this, many hours are wasted in formatting and embellishing reports and memos which are read only by people in one department.

The most common use of information systems is to accumulate data and make it easy to distribute and retrieve. The labour is in the loading. But good systems do not just pile up anything - not money, not materials and not data. They make them work. Each has its own measure of working effectively.

The measure of data working effectively is not as clear as for money or materials. Money is working when its return exceeds that of secure deposits. Materials are working when they are being converted into goods with the least effort. The laws of physics and chemistry can be used to see how close this is to the theoretical minimum effort. But what is the measure of data working? The cost of collecting it returns many times over? The Sabre system does that as does the Airline revenue optimizer. So do the banks' ATM and EFTPOS networks (as indicated by their profits). Here data works not so much by increasing productivity but capability.

The interest in productivity comes from the fact that it is a familiar part of economics. This is not so for capability. While everyone has innumerable anecdotes about the huge amounts of time and resources wasted with the latest desktop computers, including slow databases and office "productivity" tools - a hoax term if ever there was one - it is worth recalling that the "C^3I" function of computers has been fulfilled. "C^3I" is a military acronym for command, control, communications and information. In the business world this has become the ability to co-ordinate expertise and materials around the world. The Sabre airline booking system and the world-wide financial system is testimony enough to the effectiveness of computers outside the excessively hyped office environment, which is little more than an expensive typing and messaging machine. The lesson is that we should not think of information technology and information systems, the attention grabbing IT and IS. The attention should be on the slogan ASS BOTTOM: administrative system support, business operations technology and technology for operations management. Furthermore, such concepts go beyond computers and information processing, to an array of technologies, serving those purposes. Two decades is probably quite long enough to get over the novelty of relatively silly applications on desktop computers and look at what actually does things for the customers. Or even more so, how technology positions the company against competitors in potential markets.

Organisations which first understand the best use of new technology have a considerable advantage.

An historical curiosity divorced information from purpose. In the 1940s, Claude Shannon developed his theory of information which was really a theory of the most succinct coding of signals. It developed the number of "bits" idea of information. But this divorced information from its context or purpose. About the same time Harold W. Kuhn defined information as what determines which move a player in a game makes. Game theory will be discussed later but here information is linked to a purpose - advantage in a game. While computer manufacturers need to concentrate on Shannon's bits of information and attempt to define the effect of information via their idea of information, system creators, implementers and purchasers should hold onto Kuhn's concepts. Shannon's theory can be linked to concepts of disorder and the idea of entropy in physics and from there to energy, and productivity. Kuhn's ideas link to advantage, the ability to out-compete a competitor. This theoretical concept of information is closer to capability. This will be developed in later chapters.

Still, productivity sells computers. The claimed justifications of productivity occur in data-intensive administrative applications. Most of these were standardised on mainframe and mini-systems in the 1970s. Thus productivity gains in these areas, which were scattered throughout the organisation, were made early in computerisation of organisations large enough to justify the purchase of such systems. The replacement of these mainframes with cheaper equipment and sometimes fewer attendants, does not, by itself, change productivity. The lessons of this era are that it is hard to gather up increases in productivity in bits of jobs and reshuffle tasks so the gains are not simply transferred to a more leisurely pace of accomplishing other parts of a job. But there is good evidence that moving computers into operations does make a difference. In *"What's Driving the New Economy: The Benefits of*

Workplace Innovation" (NBER Working Paper No. W7479, Issued in January 2000), Sandra E. Black and Lisa M. Lynch state that, "Using a unique nationally representative sample of U.S. establishments surveyed in 1993 and 1996, we examine the relationship between workplace innovations and establishment productivity and wages. We match plant level practices with plant level productivity and wage outcomes and estimate production functions and wage equations using both cross sectional and longitudinal data. We find a positive and significant relationship between the proportion of non-managers using computers and productivity of establishments." This and other papers in the National Bureau of Economic Research emphasise that computers allied with other workplace changes do make a difference. But those other changes need to be present.

The apparent frenetic pace of change in information technology works against real judgement. It is an industry with a commercially motivated short memory. It re-invents old techniques but labels them as new wisdom; for example, all the disciplines associated with networked personal computers were established with mainframes and mini-computers. After fifty years of computers, the operating system and interface of a standard commercial personal computer could not be run on a million dollar mainframe of the previous decade. But the available business functions are much the same. Even Intranets were well established in multinational companies by the 1980s without being so named. And the information technology trade went from a trillion dollars for the decade to the end of the 1980s to a trillion dollars per year.

It is not clear how quickly things will get much better. And the promise of more amazing features is likely, according to W. W. Gibbs in an article *"Taking Computers to Task",* Scientific American, July 1997, to make computers, "more fun and engaging to use. But will they earn their keep in the workplace?" The article covers the same ground as Forester (op cit) on the productivity paradox. This would indicate that the failure of computers to provide a convincing boost in overall productivity has not yet received a convincing excuse.

> "We've spent a trillion dollars. Are we having fun yet?...It's high noon on the cost justification trail. The bean counters are closing in. What's to be done?" (Advertisement for Logical Operations in MacWeek 15/6/1992),

In the final analysis, computers are only a part of the twentieth century's systemisation and automation. One need only look at a child's book on how things are made to realise that most of the pleasures of wealth, be they matchsticks, chocolate bars, television sets, clothes, washing powder, books, sports equipment, toys and so on, are produced in large factories which can make them by the millions if not billions. The real productivity gains have been made by thousands of unsung engineers who have steadily automated the production of goods. However we want to count the wealth produced, or sucked up, by administration and services, the basis is the outpouring from farms, mines, plantations and factories. The ease with which they pour forth and distribute their goods is *the* primary measure of productivity. Information might allow groups to move the money around more quickly once it has been made. After all, the productivity of stockbrokers, or checkout counter operators does not produce any real goods, so why should it be measured as a contribution to wealth?

The United States economic boom at the end of the twentieth century was linked to computer technology. Of course the computer elite, like politicians, take as much credit as they can for the efforts of ordinary workers. Much of this was the apparent weightlessness of technology stocks which drove the United States' stock market to about 150 percent of US national income (*Brisbane Courier mail 17/5/99 p.11*). The overall stock market index numbers have weightings which have bearing on how good the numbers look. Stock markets include a weighting for commodity prices, in particular resource stocks. In 1981 resources were weighted as 62 percent of the stock market all ordinaries index. In 1987, before the October crash, resources were "down-weighted" to 25% percent. This gave speculative stocks a greater weighting of the stock market index, and a consequent levity to the market and its investors. The banks, presumably with their worth based on accumulated surplus "actual" wealth minus unpaid loans to customers, have had their contribution to the stock market index raised from 6 percent in 1987 to 22 percent in 1997. By then the Internet was the "next big thing" and produced a new peak in the stock markets' tides of crashing waves. Thus it was with the railroads in the middle of the 1800s, airline stocks in the 1920s, franchising in the 1960s, and property in the 1980s. (Curiously the period between these busts seems to be roughly half that of the previous cycle). Each boom, except property, has ridden on technologies which have been historically very important. But this does not mean they are the harbingers of every good.

The sustained growth throughout the 1990s produced the idea that economics was now entering a new phase where technology - read computer based technology - meant sustained accelerating growth. The United States had finally figured out how to use computer networks, and particularly the Internet, and a new economy was rising; one in which the "old" rules (never explicitly listed) did not apply. The evidence was a meager one to two percent increase in productivity per year throughout the 1990s, representing the ability to do in 49 days what was previously done in 50. This compares unfavourably with the 1920s overall rise in productivity of 40 percent, mostly from electrification. It compares with the depression years of about 1.6 percent increase per annum. This end of the century modest rise in productivity can be partly expected as a response of workers after a period of insecurity. The radical downsizing of corporations in the 1980s and 1990s - 500,00 to 700,000 per year with 680,000 in the boom year of 1998 - along with the business siege mentality arising from the onslaught of cheap, good quality goods from Asia, could be expected to give an earnest edge to employment. Furthermore, e-commerce, surely the most obvious contribution of the Internet to efficiency, was, at Christmas 1999, just 1.2 percent of all US trade and only 0.5 percent the year before. Even e-mail within corporations was well established before the 1990s and was much faster than Intranets. The Internet provided no explanation of productivity. The reduction in costs due to Internet business purchases requires something more than e-mail. It requires a goods commodity market; a world-wide, co-ordinated online database with a network of very efficient delivery agents.

It is not the information which is transforming economics but accumulation. The prosperity of the last half of the twentieth century, despite historically minor fluctuations, has resulted in the accumulation of money in pension funds and with professional investors, including banks and insurance companies. As noted in Chapter 1, by the early 1990s, the equivalent of $US1,000 billion moved through foreign exchange markets each day - this more than doubling during the decade. Much of it was, as always, on its way to buy stocks. And it poured into glamour stocks at a huge rate. $US7.4 billion per day flooded in from US mutual funds alone. That this did not produce an immediate inflationary effect is that it was honest money - accumulated from previous work. Spare industrial capacity also allowed the flow of goods to keep up with the available money, and so hold inflation.

Entertainment and, to a lesser extent, education, rely on the surplus of goods provided by the productive layers of society. We are now so rich, with remarkably stable and resilient systems, that mass produced entertainments are a dominating aspect of business. Entertainment / "Infotainment" companies have very fluid and vaguely seen limits to their growth. Nothing demonstrates this more than the merger of Time Warner and America Online which, on the announcement of the merger, created a company with the stock market capital of $US350 billion. The company is devoted, essentially, to leisure activities - mainly entertainment. For comparison, this world of heady values had Microsoft with a (temporary) stock market value of just under $US460 billion - roughly twice the GDP of Australia *(Brisbane Courier Mail 17/5/99 p.11)*. The number two in market capitalisation (start of March 2000) was Cisco Systems, maker of networking equipment with $US450 billion. In March 2000, it briefly took centre stage as the world's most valuable company, going over $US500 billion, before withering over the following year to about a quarter of its peak value. Cisco then slid into billion dollar losses because of excess inventory. AOL had 26 million subscribers at the time of the announcement of the merger and with a $US180 billion market valuation, was equal to twice the entire American transportation industry. Each AOL subscriber must have had a market valuation of $6,923 (How did they feel about that?). In contrast, General Electric (known simply as GE) was number three at $US435 billion (March 2000) - it soon became number one as some hint of reality reasserted itself. In 1998 the total shareholding of the largest oil company, Exxon, was $US178 billion and Mobil's shareholding was $US68 billion. The two combined to form the ExxonMobil giant with a shareholding of a "mere" $US0.25 trillion. The largest pharmaceutical conglomerate, Glaxo-Wellcome SmithKline Beecham, was created to produce a $US200 billion market valuation company with $US30 billion annual sales and a workforce of 105,000 *(Australian Financial Review 21/1/2000 p. 40)*. Finally, at the base of all economies lies resources. In March 2001 the Australian company BHP and the English group Billiton proposed to combine to form one of the largest metal and energy resource groups. Capitalisation: a piffling $US30 billion. That these figures change almost daily is just the nature of hopes and hunches.

The lessons of this fashionable enthusiasm are that the computer industry has enormous power to create its own version of what is real. It lives on cancelling history and a consumer vision of the future - one in which investors failed to realise.

Computer aficionados are entranced by interaction with computers; it is good in itself; that it is trivial is not a consideration. The Internet and E-commerce are therefore good in themselves: progressive, modern and desirable. For the spokespeople of the industry they are good because they increase the flow of products and services. The fantasy is only limited by annoying old-fashioned considerations: the cost and difficulties of moving goods around the world to any far-flung customer with a personal computer and a connection to the telephone system. But practical considerations aside, information technology was creating the "new economy." The impression is like the feeling of triumphal progress at the end of the nineteenth century, a new age has dawned. The new age of the post-industrial information economy. But unstated in this vision is that the world will become a commodity market, with local and global booms and busts. What skills best fit this new world? The best option might be to take a degree in commodities, hedging, futures and options. But have a fall back position as a hedge against being replaced by a program.

Chapter 3: From Technology to Science

The history of science is one of brilliant advances, wrong turnings, lost opportunities and the eventual accumulation of the best ideas. The equally numerous bad ideas fade into history. There is a parallel and equally important set of ideas: those of systems. Here the stroke of genius is to assemble and organise what is commonplace into something that transcends the sum of its parts. Like science its failures fade into history along with their backers. But while science grows remote from everyday life, systems embrace it more tightly.

On Science

It is commonplace to identify science and technology. Historically this is a recent phenomenon. The development of windmills, clocks, cannons, compasses, glass, steel and innumerable items of life occurred before the development of science. With the exception of glass for lenses and metal for durable, precise measuring devices, science could proceed without them. It was craftspeople and traders who, in the towns of Italy and France in the late Middle Ages, started the march of European technology. This evolving technology would colonise much of the world four to five hundred years later. The spread of technology, more than early science, did much to start an empirical cast of mind and to push back superstition . Mechanical systems - systems made from cogged wheels, levers and pulleys - cannot be made to work without attending to the components that must work well. Broken cogs, jammed pulleys, and bent levers provide ample explanation for why something does not work. The effects of confronting broken mechanical systems force the development of an empirical attitude. Invoking devils to explain why a windmill has ceased to work is stupid in the face of a broken axle. In a culture replete with supernatural explanations, mechanical systems are a powerful inducement to rely on the visible and the tactile and to discount what is not plain to see and touch. Experience refines the subtlety of explanations of phenomena. Concepts of friction, stretching, the effects of extreme weather and humidity edge the explanation of breakdowns and inefficiencies beyond the visible. Over time the empirical artisan / engineer develops a readiness to try to understand the invisible properties of matter and biology. For without this the description of mechanical systems slides into rules of thumb. Rules of thumb, when tabulated and written down, have been the basis of most technology until very recently.

Building, perhaps more than anything else, illustrates the difference between science and technology. Building was our earliest advanced technology. It is what we take as the measure of civilisation. The bigger and more elaborate the buildings, the more advanced the civilisation. This could be taken as the archaeologists' bias - big buildings survive the ravages of time and so tell us more about ancient civilisations. But more than this, they tell us that the society was organised to concentrate resources, usually wastefully, in tombs or monuments of glorification.

I met a traveller from an antique land
Who said: Two vast and trunkless legs of stone
Stand in the desert. Near them on the sand,
Half sunk, a shatter'd visage lies, whose frown
And wrinkled lip and sneer of cold command
Tell its sculptor well those passions read
Which yet survive, stamp'd on these lifeless things,
The hand that mock'd them and the heart that fed;
And on the pedestal these words appear:
'My name is Ozymandias, king of kings:
Look on my works, ye Mighty, and despair!'
Nothing beside remain. Round the decay
Of that colossal wreck, boundless and bare,

The lone and level sands stretch far away.

Percy Bysshe Shelley

Shelley takes poetic licence on the Tomb of Ozymandias which is really the funerary tomb of Ramses II. The Egyptian builders were probably the first to develop geometry and surveying and a sense of precision. Inevitably they learned a large number of rules of thumb from collapsing pyramids and other failed monumental structures. They started the tradition of competitive glorification. First came the glorification of kings, then emperors and, most enduring, states and gods. This has provided humanity with many of its most admired public creations. The Parthenon in Athens is a subtle fusion of precision geometry and architecture which symbolised the wealth and power and intellectual prowess of Athens during its brief burst of glory. The Taj Mahal, the tomb of Nurjahan, the wife of the Mughal emperor Shahjahan, is love expressed in architecture and the world's most admired tomb.

By the time of the Romans and the Chin Dynasty in China, huge building projects for the public good were normal. Roadways, aqueducts, hydraulic systems for irrigation are huge building projects consuming social energies as much as the temples and palaces. Unspectacular though these might be, they are often remarkable achievements. High buildings with vaulted ceilings and domes require planning, experiment and calculation. This is technology: specialised knowledge handed down within the trade. It is building technology rather than the buildings themselves which must have been a fertile field of simple, though possibly large, systems. The Medieval architect and engineer Villard de Honnecourt, famous for his mid-thirteenth century notebooks, lists as building machines hoisting machines, engines for raising water, water wheels, water mills, the water screw, various pumps, water organs, the hodometer and catapults. None of these could have been created without tables of sizes of components. The Flying Buttress , the key invention for Gothic cathedrals, would have had its table of components and their requisite sizes. These would have been the result of accumulated experience. The concepts of force, stress, load, and strain were still to be clarified. Geometry was the art of geometry - various techniques for drawing and locating points or lines. This of course is really a late entry in the world of building technology. The Chinese had their own architectural geniuses and tools. The Islamic empire, from about 800AD to the 1500s, covering an area which went from India to Spain, was much better organised than Europe and had its own architectural achievements and no doubt its own tools of trade developed by its architects of genius.

The knowledge we have now of stresses and strains on beams and supports issues from the age of steel construction. The great cathedrals with high domes, built to last millennia, were built by rules of thumb. The technology was not so complex as to prevent Michelangelo designing the Dome of St. Peter's in his eighties. Genius though he was, he was not known to be a mathematician or geometer. Had the technology demanded such specialised knowledge, he would surely have recorded his travails, for he was not shy about complaining. The Dome of St Paul's in London was designed by Christopher Wren, known as a mathematician but no mathematical contributions came from that work. From Japan to Britain building technology has produced spectacular results but without the help of science.

Buildings are static. They do not do anything, they transform nothing and so they are not a measure of a society's grasp of systems. Nothing comes from the building except a feeling of satisfaction and wages.

Prosperity needs systems thinking more than scientific thinking . But without scientific thinking the improvements become a matter of chance.

This is illustrated by moving east and west and back again.

The West has only recently become aware of its debt to China. For 1,100 years between 300 AD and 1400 AD, China was undoubtedly the greatest nation in the world. Nothing was comparable to it. Its power and technology, agriculture and political systems, when operating well, gave its rulers unmatched control and ability to organise an opulent aristocracy. The list of technology is extensive. It starts with an efficient agriculture. Growing crops in rows, hoeing and the iron plough gave China a productive agriculture which allowed it to feed its people more effectively than the West. These innovations were all in place by the fifth century BC. The seed drill, fishing reel, stirrup and wheelbarrow are all simple but important inventions. By the first century BC, bamboo pipes to pipe natural gas drilled from 4,800 feet down were being used. Steel from cast iron was available at this time. Paper was used for clothing, writing, armour and money, and the segmental arched bridge, belt drives and chain pumps were all developed before 1000AD. Wood block printing was developed more than 300 years before it was in Europe and there is some evidence that Gutenberg knew of the idea before his 1458 invention of moveable type. Gunpowder and the magnet are the well known Chinese inventions. The circadian rhythm and the circulation of the blood were "normal" ideas 1,600 years before they were common in the West. The technology of junks as reliable long-distance ships is in a class of its own. From 1100 AD onwards, they were used for sea trade ranging from India to Java. From 1405 to 1433 Admiral Chen Ho (or Zheng He) used fleets of up to 62 junks, in six different expeditions, to explore the African East coast as far South as Madagascar. These were not just round the coast ventures but ocean going, compass navigated, explorations. China could have become the dominant and enduring world power from then on. But instead it closed down. The Middle Kingdom of the Ming Dynasty required nothing of others. It decided it was perfect, knew everything and chose to compile existing knowledge. This produced a 370 volume 11,000 chapter encyclopaedia completed by 1410 and *The Great Pharmacopoeia* by Li Shih-Chen which appeared about 1596 and which discussed smallpox inoculation.

For all this it remains a mystery why China did not invent science even before the late fifteenth century. Some scholars have argued that the Judeo-Christian tradition of a law giving God makes it permissible to search for the laws He used to create the world. This gives the Arab nations an equal chance as well. The argument goes that the millions of Gods in the Hindu tradition and the formless unnameable aspect of Taoism in Chinese thought, formed a social barrier as well as an intellectual hurdle in India and China.

The Islamic Empire was well positioned, sufficiently organised, more civilised and tolerant than Europe was until about the 17th century. Its religion is the same monotheistic cast as the Judeo Christian tradition; indeed they would describe it as the culmination of the Judeo-Christian line of development. By the above argument, if this is what it takes to produce science, then they should have. Which of course they did. The Muslims where crucial in mathematics. Al Khwarizmi, who lived in the first half of the 9th century and flourished in Baghdad's "House of Knowledge", was the first non-Indian to take up "Hindu" numerals. He wrote a number of works which developed algebra. His transposition (*al-jabr* in Arabic) method of solving equations gave us the word algebra. His name is immortalised in the term algorithm. He and other Muslims measured the earth's diameter with remarkable accuracy. They realised that accurate results required accurate instruments and set about to create them. The Arab use of Indian numerals transmitted them to Europe where Leonard Fibonacci, a merchant in Pisa, wrote a treatise which showed they were a better way to do arithmetic. This occurred in 1202. After some hesitation, the Italian merchants were convinced.

As with China, the Islamic states anticipated the science that was to be rediscovered in Europe. Around 1000 AD Ibn al-Haytham had worked out much of optics and the nature of the eye.   In the 13th century Ibn al-Nafis had described the pulmonary circulation of the blood, anticipating the Europeans by about 400 years.

The idea that science was born in a particular culture is an anathema to post modernists and other sociologists for whom science is merely a useful ideology of control for powerful elites. It would be argued that of course India, Islam and China had science; to deny them science is to narrow the definition to give ownership to the West.   Certainly they had the elements of science; they had good measuring instruments; they had rigorous thinkers and many achievements.   They had the desire for power which often drives new thinking.   Perhaps a characteristic of science is the rate at which thinking is sustained and communicated.   This means book publishing, which the Chinese had, and eventually Europe had.   But with the normal understanding of word in the East as well as the West, the science that has its greatest impact on our prosperity, the science that counts in manufacturing, and the science which counts in organising resources, occurred in Europe.   The first crucial idea, but not one which was unique, was the Greek penchant for geometry.

To construct a permanent rigid building requires planning and measuring.   It requires a sense of geometry.   Geometric propositions and studies are world-wide.   But the Greeks made the subject their own.   They developed the famous axiomatic deductive study of geometry.   Euclid (circa 300 BC) and Pythagoras (circa 500 BC) are justly famous as pioneering theorem provers. What they proved was known to many in other nations but they can claim to have *proved* things.   The *style* of thinking is what counts. Above all this sense of rigour survived the degeneration of Europe into ignorance and Feudalism between about 400 AD and 1300 AD.   Much of the Greek learning, which became the province of the Arabs, had diffused back into Europe by the 13th century.   By that time European architects and clockmakers were ready for a bit of mathematical help.   Even burgeoning banking operations, not to mention a bit of gambling, were posing mathematical problems all of which could do with help from any direction. But it was in the hands of Isaac Newton in the 1660s that Greek rigour finally fertilised the study of nature, and science, as we know it, was born. But as with de Vaucanson, genius vanishes if the conditions are not right. Newton's insights had to be part of a journey already begun.

One of the most important steps started with Nicolaus Copernicus . He was the first major modern European to re-emphasise the idea of a sun-centred solar system. The idea was published in 1543 when Copernicus was on his deathbed. Johannes Kepler (1571 - 1630), working over the measurements of Tycho Brahe (1546 - 1601) and applying them to the sun-centred solar system produced the first astronomical law: that the journey of each planet around the sun is an ellipse and the time taken to move part of the way around the ellipse is directly proportional to the area enclosed by the sun as one of the focal points of the ellipse and the points on the ellipse that mark where the planet started and finished. This means that if the planet is travelling fastest close to the sun, it sweeps out the same area for a one day journey as it does for a one day journey when it is furthest from the sun and moving most slowly. This was a triumph of observation and geometry over the prejudice - held for a long time by Kepler himself - that the planets revolved on celestial spheres. A string of discoveries followed. Gallileo (1564-1642) is a crucial figure in astronomy with his pioneering use of the telescope, the discoveries of which forced the acceptance of the Copernican model of the solar system. He also made significant contributions to the concept of inertia and the properties of gravity. Sir William Harvey (1578-1657) pioneered Western studies of blood circulation and embryology. Robert Boyle (1627-1691) initiated studies which moved alchemy towards modern chemistry. The towering genius was Isaac Newton (1642 - 1727).

This taciturn, unsociable genius, who spent a majority of his intellectual efforts dating events in the Bible and doing alchemy, understood very clearly the core of science. You can only check things when you have something very clear to check. And the clearest of things to check is a measurement. But for something to be new knowledge, that measurement must be predicted from a mathematical description of other phenomena. His statement of the nature of gravity was just such a triumph. The statement is not at all complicated: that the attractive force between any two objects can be found by multiplying their masses together, dividing by the square of the distance between them and then multiplying by a number called the Gravitation constant. This number adjusts the attracting force according to the scale used to measure masses and distances. (In the normal scale of measurements, say kilograms and metres, the Gravitation constant is very small, reflecting the fact that we can easily lift a kilogram weight against the attractive force of the whole Earth). This law, with its succinct equation, gave rise to a host of dynamical "Universal Laws".

The very small number of previous laws, such as a Archimedes' Principle (that a body immersed in a fluid will displace a volume of fluid which weighs as much as the body would weigh in air) were "static". Dynamic laws, which describe changes dependent on time, are much more important because simple changes are easy to observe and measure, especially if you have good clocks - something which Europe did have. With the law of gravity, Newton could calculate the behaviour of planets. He could derive Kepler's laws from a much more general proposition. It was also why calculating the position of the moon in the sky gave him headaches (literally). Taking into account the influence of the sun's as well as the earth's gravity on the moon, made the mathematics extremely difficult. Especially when he was calculating numbers which had to distinguish between his theory and well tabulated rules of thumb. But the principle stood. Newton knew that you haven't done anything unless you have provided a number which can be checked. If an investigation of the phenomenon gives a different number, then the theory is wrong in some way. Back to the drawing board. Oddly enough, this principle, a routine practice in the "hard" sciences of physics and later chemistry, wasn't really clearly stated among those who pontificated on science - the philosophers of science - until Karl Popper (1902-1994) called it the *criteria of falsification* in 1934.

Popper gave his widely accepted *prescription* of science by discounting the verifiability of theories from observations. Science, he declared, proceeds by bold hypotheses - imaginative theories - which are not at all verified by observation so much as rejected after conflicting with observation. Popper claimed that a crucial aspect of a good scientific theory is that it should be *falsifiable*. There should be some observation that can be made which could conceivably turn out to contradict the theory. If the theory is false, then we could see that it was so; in short, the theory should make some difference to the way the world is. The crucial aspect of this is to frame hypotheses in ways which can be checked. This is done by insisting that the ultimate ideas must be reduced to things which can be counted or measured. Furthermore what can be counted or measured can be done anywhere and anytime by anyone who has the right equipment. The last requirement can be a problem but doesn't destroy the objectivity of science. It makes science public, objective, and non-mysterious. It also makes a good filter for rejecting the poorly described, the dubious and the slippery.

For example, any postulated entities such as ghosts or spirits which live on a different "plane" and do not interact with this world cannot be the subject of a scientific theory - there are no conceivable observations which can falsify the theory.

This brings up the nature of observation: what counts as an observation that can be compared with theory. Consider the theory of "body language" in which hidden emotions can be revealed by decoding a person's "body language." What counts as a defensive or submissive body language? When does it begin and end in time? How often do you expect such body language; always, 90 percent of the time the person is feeling threatened or whatever, 25 percent of the time? Rigour demands that there is a specified number, or count, which should be observed a specified number of times. This count or measurement can be a complex measurement - it can be the rate at which a population increases or decreases or even the rate at which the rate changes. The measurement, or count, can then be compared with the theory and one can clearly say "yes, this is in the bounds of the theory" or "no, this not what the theory predicts". The number is wrong or the theory is wrong.

The two most important things about properties or measures which are to be useful in science, are that they are definite: they come back with a precise number or a straight forward answer "yes what we are looking at is a member of a particular set" or "no it is not". Nothing lies in a vague half-world of maybes. Body language, in its popular statement, is not definite enough. Another requirement of science is that a property or measurement should be "effectively decidable". There must be some way I can decide in a reasonable number of steps that a property is true or that something is definitely a member of a given set. The way a measurement is made can be given by a set of instructions made up from steps which have proved to be achievable.

This requirement is not satisfied by much of what is taken to be science. For example, a typical principle of psychoanalysis is that the subconscious is the repository of painful memories which are censored from the conscious mind. Only the intervention of a psychoanalysis can overcome the power of the subconscious censor. How are we to identify whether a "memory" which is reported during a psychoanalytical session was one dredged up from the subconscious - and which dodged the censor - or was a previously censored memory, or was just one of those "that reminds me" events? How definite are the properties used in this statement? How do we go about deciding that a memory is painful? That it has been censored?

The life and human sciences complicate Popper's theory by dealing with statistical measures and distributions. It can be hard to discover whether a measured distribution is distorted by a sampling error or whether it really does conflict with a sociological theory. Worse still are assertions which are just statistical correlations. The observation that boys from homes with ineffective or absent father figures are a "high risk" for crime, is a statistical correlation. It is not much use in telling you what will happen in individual cases. Probability distributions and statistical correlations, while alerting us to some connection, are weak science. Exposing them as false can require repeating studies almost as large as the ones used to calculate the distribution or correlation. A better approach is to have dynamic theories which predict changes in distributions. These give a greater number of things to compare and so testing the theory can be a complex project. However, changes involve a greater variety of phenomena and so contain more information which might reveal flaws in the theory. But it can be done. The very foundations of modern genetics are proof of this. It is science of the highest rank which came from an unexpected place.

The obvious observations of the way children mix the characteristics of their parents are known to all humans. The domestication of animals was, in some sense, a breeding program for the characteristics of "domesticity." The scientific study of genetics was the idea of Gregor (Johann) Mendel, the Abbot of the Augustine Monastery at Brunn, in what used to be Moravia (now part of the Czech Republic). His experiments in "plant hybridization" were done in his spare time in the monastery garden. Mendel theorised that inherited characteristics were discrete - that is identifiable in units and the distribution of these characteristics was caused by the pairing of elementary units of inheritance, now known as genes.

Mendel crossed varieties of the garden pea which had previously been observed to maintain constant characteristics.

The crucial step that Mendel took was to make a list of characteristics which were clearly in the pea or absent from it; he checked for tallness and dwarfness, presence or absence of colours in their blossoms and axils of the leaves, and similar alternative differences in seed colour, seed shape, the position of flowers on the stem and the form of the pods. He was able to count all these characteristics in the various hybrids and used what are now standard statistical methods to give expression to his hypothesis of units of heritability. At the time, about 1860, the use of probability theory techniques such as binomial distributions was very rare, and it took some time for the brilliance of his insights to be appreciated.

Mendel's work is a clear example of the idea of a theory composed of the combination of abstract elements - namely, inheritable units, or genes, which have rules of mixing - and these abstractions giving rise to phenomena which can be counted or measured. Mendel's published results are a fudge - they are too good! Nevertheless his clear insight on what it takes to make something scientific makes him a scientist of the highest rank. The addition of Mendel's genetics to Darwin's insights - sometimes called Neo-Darwinism - strengthened the scientific content of the theory of evolution so that it is now the cornerstone of modern biology.

In some ways Mendel's work is on a more abstract level than that of physics. What he was measuring was the change in the distribution of things which could be counted. Just counting traits on their own did not amount to anything. The theory was about the way things mixed and redistributed themselves. No description of the mechanism is required - that came about ninety years later with Crick and Watson's famous unravelling of the structure of DNA. Changes in distributions are more abstract than the position of a planet or the temperature or pressure of a gas. It has to reduce large amounts of information to simple measurements from which a comparison can be made. Mendel's work had nothing ostensibly complicated. No "rocket science" there. Just a high level of abstraction.

As with Mendel's work, observations are usually observations of the way things change. Even theories of how things are "at equilibrium" must have something to say about change close to the equilibrium state. If they do not it is often hard to come up with a good experiment to test the theory.

Popper's prescription implies that science is not so much verified but winnowed. The good scientific theories are those rich in predictions which have been tested and not found to be in conflict with the observations and experiments. This is similar to the evolution of systems. Here the criterion is much more pragmatic: we keep modifying systems when they do not adequately process the inputs we throw at them. Similarly, a scientific theory has to process inputs in a way which accords with its specification: that it predicts another set of measurements. To do this it must be much more precise that statistical correlations. These are useless when trying to produce a system.

Mendel's work illustrates another aspect of science. It is not the same as technology. Science does need technology for much of its development. Newton and Einstein gave us classical physics and gravitation and the intellectual tools to start our understanding of astronomy, but they neither relied on, nor directly produced, any technology. The same is true also of Darwin's work. But it is technology that gives science its usefulness and public creditability. Today the march of science is enriched by technology and, in turn, enriches it. Three illustrations will suffice.

It is well known that air-conditioning costs as much as heating. If you live in a very hot or very cold climate and want to remain comfortable the electricity utilities are the winners. The explanation for this arose early in the industrial revolution. A simple question for an early textile mill operator was, "How much fuel is required to produce a yard of cloth?" The understanding of the conversion of heat to work was not at all obvious. Newcomen and Watt had to face the question squarely.

But it was Sadi Carnot (1796 - 1832), who wanted to know why French steam engines compared so poorly with English ones, and who produced the first significant advance. In his *Reflections on the Motive power of Fire* (1824), he had the right idea from the wrong reasoning. He thought of heat as an indestructible "fluid' and reasoned it did work by dropping in temperature. He proposed that no engine could be more efficient than a reversible one (which is true - it would be 100 percent efficient). (This principle has its counterpart in Landau's principle: a computation does not use energy if it is completely reversible.) Carnot's work was extended by Rudolf Clausius (1822 - 1888) and William Thompson (1824 - 1907), who became Lord Kelvin. Kelvin backed the idea of James Joule (who gave us the alternative energy measure to the calories listed on food packets) that heat was a type of motion. Kelvin reformulated Carnot's work in terms of the motion theory, and in doing so, added the concept of absolute temperature, whereby at absolute zero there is no motion, and the impossibility of transferring heat from a colder to a hotter material. Clausius, with similar ideas, took Kelvin's ideas and developed the crucial idea of entropy which is a measure of disorder or unusable energy. From this was developed the second law of thermodynamics (crudely: concentrations of energy disperse and do not spontaneously concentrate). All of these ideas were put into their modern statistical form visualising heat as atoms or molecules in motion by Ludwig Boltzmann (1844 - 1906).

In 1852 Kelvin and Joule measured the fall in temperature of gases expanding from a nozzle. The fall in temperature comes from work done as the motion of the molecules expands the gases. The observation is commonplace; blowing air from an open mouth is warmer than when the lips are pursed. The measurement and explanation is profound. This phenomenon led to the liquefaction of gases and eventually the discovery of superconductors. But that was later.

Meanwhile as these ideas circulated…

In 1823, the greatest experimental physicist of the nineteenth century, Michael Faraday, noticed that when ammonia condensed in a metal tube it produced intense cold. Curiously he did not exploit this observation. This was left to Ferdinand Carre in 1859. Carre produced the first widely used refrigerator which used ammonia as the refrigerant. He had been preceded in the United States in 1844, by John Gorrie who used water to cool compressed air, which, on expanding, cooled sufficiently to create ice. Also in the United States, in 1856, Alexander Twinning had used a vapour compression technique to supply ice. The stage was set for a revolution in food distribution. The refrigerated rail allowed the concentration of meatworks away from their markets which added to the growth of Chicago as a meatworks town. The idea was quickly adopted for international transport, starting with the French ship *Frigorifique* in 1876. Refrigerator ships were the tools which grew the economies of Argentina, Australia and New Zealand.

Electricity replaced steam as the motive power of choice in the first part of the twentieth century. This was the first technology derived from science. The discovery that an electric current is produced when a conductor is subjected to a changing magnetic field is not something that comes from working with ordinary materials. The effect is too small. It was investigated by Joule at the start of the 1840s. He quantified the link between heat, resistance and current. Joule then established the ratio of mechanical energy to heat and provided the link between mechanical work and electricity. In 1867 Charles Wheatstone in the United States and Werner von Siemens in Germany, independently applied these ideas to produce the dynamo. Now heat from coal or motion from falling water could produce electricity which could be transported via wires to where it could in turn be converted into mechanical power or for heating and cooling. With thermodynamics and electricity, science moved away from the mechanical explanations to the subtle properties of matter and produced a new world. This is a world of distributed energy and multiple forms of energy.

And this is the practical guts of science. Never mind the hopeless muddying of the meaning of the term by various philosophers and sociologists who play games with the term. In spite of the overwhelming evidence for the pillars of modern science: atoms with all their subtle structure, the micro, chemical and cellular analysis of living creatures, there are schools of philosophy and sociology which see this as a Western "myth". This myth is claimed to be propagated by various power structures for their own ends. The ends are not clearly linked to any actual alleged mythical content of science. Perhaps it is to assign a rationale behind the media / commercial / governmental / military complex which depends on modern technology for the continued good working of the power system. A popular, if verbose, version of this is postmodernism, for a while the dominant fashion in university humanities departments at the end of the 20th century. Whichever perspective you want to take, Popperian or postmodernist, the theory that science is a western power myth is not, itself, science. Curiously, for postmodernists, statements about science have a more certain status than the actual statements of science. (One wonders what they might do if left to play in a chemistry laboratory). But science as described above is what has accelerated the flow of goods out of factories. It is this narrow, practical concept of science that has produced flight, electronics, the modern media, and the pharmaceutical industry which underpins modern western medicine. And it is the apparent fruits of technology underpinned by science that have tremendous force for intellectual conversion if not colonisation.

By the end of the nineteenth century science had proved itself useful in business. The large science-based industries of the twentieth century were only just getting started. Photography, dyes, pharmaceuticals and the Bessemer process for manufacturing steel were already established. From the perspective of the present century, the marriage of science and business is overwhelming and obvious. In the nineteenth century it was a minor social phenomenon because it involved only a small number of people. But the effect of those people was enormous. The change in ideas over that century and certainly from 1810 to about 1910 was enormous. It included large-scale manufacturing along with many management techniques, the establishment of the foundations of chemistry (the periodic table), genetics, evolution, anaesthetics, immunisation, electricity, radio, X-rays, the first glimmering of atomic structure, refrigeration, flight and automobiles. Including Babbage's computer, we can even add computers. Extending this time to 1914 the skies were well and truly conquered. Eleven years after the Wright Brothers' first flights, Sikorsky's *Ilya Mouremetz* stayed airborne for six and a half hours and carried six passengers. The twentieth century elaborated and refined this list, adding - for general benefit — plastics, antibiotics, quantum mechanics (giving us the transistor and electronics), and the biochemical understanding of life. Much of twentieth century science, building on the achievements of the nineteenth century, is arcane and complex, although almost all of it has found an application.

There is also a new dimension to science which started to creep into the last two decades of the twentieth century. This is the contribution of computers. It is the use of the computer to display data which demonstrates new patterns, and to create algorithms to describe interactions. This is sometimes called 'phenomenology'. It is the recognition that a pattern or a regularity exists and should be investigated. The regularity can be startlingly clear but there is no theory to explain it. A great deal of what makes us a perceiving animal lies in this area. How do we, or even our pets, recognise faces and sounds? More generally, there are people who are obviously more expert than others at recognising things and judging them as requiring a particular type of response. While not proceeding from known scientific theories, their knowledge does belong to a realm of systematic knowledge which must have some underlying logic. This knowledge is part of the phenomenology of a "problem domain". This means these experts have the rules of thumb for giving a decision where there is no algorithm or "theory" behind the decision. Frequently the rules are the familiar lists of "in these circumstances I do such and such". Not all of them are based on rules of experience. There are some, usually in finance and stock or commodity trading, but also in medical diagnosis, which do arise from complex theories. Turning the rules of thumb, the heuristics, and the fragments of theory into algorithms creates an "Expert system".

Expert systems based on rules can usually be put in terms of yes / no answers to sequences of questions. Recognition problems frequently cannot. Some progress has been made in these types of problems using the so called "Artificial Neural Networks " or ANNs. ANNs are not computers in the usual sense. They are best described as devices which take a series of incoming signals, filter the signals by testing whether a certain component is above or below a significant threshold and, if it is, "weighting" the signal. This determines how important the component is. The entire series of weights is then added and a decision is given according to this final sum. The output is frequently a simple yes or no. This result can be fed into an ordinary computer for further processing. ANNs have been successful in recognising specific faces, determining whether phrases are important to summarise journal articles, and generally detecting patterns in all kinds of biological, astronomical and battlefield data. Linked to a chemical analyser, an ANN can be trained to recognise low concentrations of chemicals such as nerve gases. The catch is that the thresholds and weights have to be determined by trial and error. The ANN is presented with the patterns and, starting with a random assignment of thresholds and weights, is told whether the attempted answer is true or not. After many trials it arrives at an answer. A variety of techniques helps the ANN, during a number of "learning" trials, to use its errors to arrive at a set of thresholds and weights which reliably gives the right answer. The thresholds and weights can be "read off" to form a single mathematical expression which can be used in a computer program as a substitute for the ANN. A related technology, genetic algorithms, does a similar task by varying algorithms but its teacher is built into it through its targeted evolutionary "gradient". Artificial Neural Networks and genetic algorithms are two types of discovery processes for finding algorithms to be used in a computer program. Their importance lies in the nature of the discovery. It is not done by understanding the nature of interactions to begin with. It works by using a computer to sift through large numbers of options to arrive at an algorithm which "does the trick". It is like solving mathematics problems by working from the answers at the back of the book. Comprehension and insight sometimes follow, and sometimes understanding is confounded.

What has this to do with science? Most science has to do with interactions in nature. Perception and recognition are part of nature. Artificial Neural Networks can give insights into what the minimum complexity of a system is in order to recognise something. That is a contribution to science. Working out the logic of how an expert guesses the movements of commodity prices might be a lead to mapping out the most important interactions which govern these movements. These systems then are part of the discovery of new interactions. They might not give us the actual theory but they might tell us how much data or how many factors will be needed in the underlying theory - or they might not. An expert system which followed the logic of the Japanese or Roman sword makers would capture the technology and some of the technique, but the principles of metallurgy would remain hidden.

Japan: Catching up on Science and Technology

At the end of the nineteenth century the Japanese had very good reasons to appraise their own productivity and technology in comparison with the West. They had been humiliated by the

US Navy and were acutely aware that their self-imposed isolation had left them weak in a world where the nations they had shut out for so long acted like unprincipled bullies.

Their determination to catch up led to a government backed study of Western science and techniques. It was clear to them that science underpinned Western military technology. It was also clear that standardisation and expectations of high quality were part of the productive power of the West. This story is told in detail by Tessa Morris-Suzuki in "The Technological Transformation of Japan".

Japanese artisans and entrepreneurs, with no formal technical training, had repeated, in the late 1800s and early 1900s, what similarly endowed artisans had accomplished 50 or so years previously in Europe and the United States. This is a slow way to play catch up. A number of academic, government and private organisations created "research institutes" to learn from the West and innovate for themselves. Japan had an estimated 700 university and technical college graduates in 1900. This increased to about 2,500 in ten years. These institutes were not exactly a powerhouse of innovation so much as a channel for transferring Western technology. And they were few and far between in industry. Hitachi recruited five scientists between 1911 and 1912. Twenty years later, Tokyo Electric's Mazda Lamp Laboratory employed about 300 trained technicians. In the United States, Bell Laboratories had over twelve times that number and Dupont four times as many. As if to prove the point that technical people are a good investment, even with this skill imbalance, Japan was the third largest consumer of electricity by 1936.

What were all these researchers doing? What new discoveries issued from all these industrial laboratories? The newsworthy part of science is new products: radios, light-bulbs, drugs, nylons, carbon-fibres and so on. But science has a crucial, though humble, role to play in perfecting manufacturing systems. As noted before, high-volume manufacturing is de-skilled, rule-based, "no-brain" manufacturing. This can only be achieved when manufacturing's raw materials are consistent and the processes well understood and predictable. Science is applied to secure standard materials for high-volume manufacturing. And we all expect it. If electric cords had wire which broke when bent, or too much insulation, or insulation which flaked off, we would be infuriated. Just about every item of household goods is manufactured with extraordinary consistency. The paper in books and newspapers, fabrics and their dyes, crockery, cutlery, glass in windows, electrical goods, ball-point pens, detergents, plumbing, the water we drink (at least in the cities where I have lived), the steel in our cars and the petrol in their tanks are all of very high and remarkably consistent quality. Vast flows of goods have had armies of scientists or science-trained technologists standardising materials fed into manufacturing systems so that these systems can turn the standard inputs into the standard outputs.

Before the Second World War Japan had to struggle with systemisation. It had to invent corporate cultures with rule-abiding loyal workers. It had to educate its workers from making rickshaws with twenty or thirty components to making cars from several thousand components. Technology-based firms had to struggle with the variable quality control of their suppliers. The quality control revolution after the war was just a scaling up for the international market of what was faced before the war. In some ways the achievements are extraordinary given the ability to produce the degree of innovation and technical sophistication which surprised the United States from the time of Pearl Harbour. The most prestigious concentration of expertise before the Second World War was the Institute for Physical and Chemical Research known as Riken (as an abbreviation of Rikagaku Kenkyujo), directed by Okochi Masatoshi. Okochi recognised that war was no longer a question of military might (and, one might add, lunatic fanaticism) but in 1937 observed, "Future wars will be wars involving the entire nation's scientific knowledge and industrial capacity." Riken, a government

101

initiative, focused on machine tools and industrial machinery but also had laboratories that did fundamental physics. The calibre of the people in Riken was very high. At one stage it included Tomonaga Sin-itiro who sprang a surprise on the United States after the war when it was found that a group led by this native-trained physicist had equalled the US effort to solve fundamental and extremely difficult theoretical problems in physics. For this he shared the 1963 Nobel Prize in physics with the famous US physicists Richard Feynman and Julian Schwinger.

The economic and sociological explanations for Japanese and German recovery after the Second World War include many factors. Two common ones were that they were recipients of the most generous acts in history: the huge American aid effort - largely to stop them going Communist. This allowed factories to be built from the most modern technology of the time. Finally, by demilitarising Japan and Germany, both countries could focus their mastery of technology on building consumer products. Both countries emerged from desperate ruined societies, stronger than before. Japan rose to pre-eminence in steel, shipbuilding, cars and electronics. By 1972 Japan had concentrated much of its research in industry, being second only to the United States in industry-employed scientists. Thus the gap between research and production was small. Germany did much the same.

Germany started with a reputation for quality. Following the suggestions of the American business consultant W. Edwards Deming the Japanese developed quality circles and eventually rivalled Germany with its reputation for quality. It popularised manufacturing system techniques such as Just In Time and Kanban manufacturing. These techniques arrange for components to arrive in small batches to be at hand for the workers as required rather than being stored in large warehouses somewhere in the factory precincts. The recording of stock used and required in the factory is then controlled by a simple card system. Of course all this was known to other companies but in a small tightly knit country it is easier to manage. The quality circle provided the insight that the system must check at each stage its output, not releasing components until they are of sufficient quality. This stops effort being expended on something that will inevitably be rejected. It is a fundamental aspect of large scale systems.

Systems Science - Science of Systems

The Japanese mastery of technology has made them experts in the integration of systems. They can now claim they can teach other people the science of manufacturing. In what ways is manufacturing a science? It is really part of a new level of science which might be called systems science. The science described by Popper is really the study of interactions of nature. They are the raw material of the logic of natural systems. Natural systems such as a star, the weather, ecosystems or animals, are huge collections of matter or organisms all with their specific interactions. Working out the interactions of various groupings of matter or animals has been the triumph of "interaction" science. The description of large-scale phenomena which arise from huge numbers of individual interactions will be the science of this century. And it will become a dominant aspect of systemisation.

Systems contain lots of boring simple bits and pieces or they are made up of simple activities and transactions. Each seemingly well understood on its own - like bank transactions, or flushing a lavatory, turning on the heater on a cold morning, driving a car to work. Repeat these over many, many people and the way we have to think is in statistical terms: the change of distributions. The individual bank transactions have to be whizzed through the system in some order. The payments in and payments out have to be tracked so that accounts don't dip into the red and reject payments which would have been accepted had the sequence been right. The statistical peaks of flushing toilets during ad-breaks on television have threatened to overload sewerage systems. The effects of rush-hour traffic are well known. Turning on heaters in a sudden unseasonably cold day can overload regional power grids. And more subtly, events in nature which are normally unnoticed can affect systems - as did solar storms affect electric power grids in the late 1980s. Electromagnetic energy from the sun can send huge waves of electrical energy surging along power lines, shorting circuits and burning out equipment. The solar storms shower the earth with electrically charged particles. When these magnetic bursts hit the earth they can cause power outages and disrupt satellite communications. In 1989 just such a pulse struck the earth and completely knocked out the electrical grid in parts of Canada and the north-eastern United States, leaving more than eight million people without power for days. This solar storm also caused the coils in a transformer station in Salem, New Jersey to melt and catch fire, causing a regional blackout. System science is about the effects of many, many simple actions and how those actions depend on stable conditions.

A system can be looked upon as a hypothesis that a way of doing things will fulfil a purpose. Daimler and Benz's first internal combustion engine was based on the crazy idea of using exploding petrol to drive a piston. Why not stick with gas or use puffs of gunpowder? The motorcycle, and then the motor car, driven by a petrol engine (in 1885) started as "just an idea". The Wright brothers' first aeroplane was based on the hypothesis that this was the sort of contraption which would fly. A great deal of experiment and theorising went into that hypothesis, which could have been wrong. Henry Bessemer's steel making process utilised a device which was just a good guess. The tilting converter was a great success when introduced in 1856 until it turned out that it only worked for phosphorous-free iron. But the experiments had been done and the device found to be useful, and was a valuable contribution to the supply of wire for telegraphs. Further refinements, which took over twenty years, overcame the phosphorous problem. Whenever innovations occur, whether they are a new toaster, a microwave oven, a submarine or a sewing machine, they all start with the hypothesis that this device will fulfil the intentions. And there are measurements which announce the experiment a success or failure. The Wright brothers had a goal of the length of the flight. Richard Pearse flying his monoplane in the first half of 1903 in New Zealand, did not claim success for his experiments because he wanted to turn during flight (something that the Wright brothers did not achieve until 1905). Bessemer had only to show that his process was a nominal percentage more efficient than competing processes to claim success.

If devices, artefacts and processes can be regarded as experiments to confirm hypotheses, then an entire manufacturing system is the same only bigger. They have had their failed experiments - the manufacturing disasters which have vanished in humiliation and lost money. The process is rational, empirical and has its hypotheses. Successful experiments are transmitted as "good practice" but underneath this good practice is a logical model. Good practice means the appropriate use of quality assurance techniques. When to monitor automatically, when to use robots, where to give responsibility to workers. This is little more than the scaling up of production engineering. The goods to be produced are manufacturing systems.

It is not only manufacturing organisations that are experiments. Experiments also occur in other activities, especially finance. A spectacular failed experiment was the hedge fund Long Term Capital Management. It was set up in 1994 by the very successful bond trader John Meriwether who invited Professors Merton and Scholes to become partners. Merton and Scholes' work with Fischer Black on option pricing won them the Nobel Memorial Prize for Economics (Fischer Black had just died). The firm was spectacularly successful at first, returning 20 percent and then 40 percent on investments. This looked like a complete validation of the partners' methods and models. However, first the 1997 Asian financial crisis and then the Russian refusal to honour its loans left LTCM exposed to the tune of trillions of dollars. Even as it was being publicly bailed to the amount of 3.5 billion dollars, papers were being published showing how the models they were using were based on stable probabilities. The market had falsified the hypothesis which was the basis of the experiment. It had worked as long as the probabilities in the formulas were unchanging. But the various crises had invalidated that hypothesis. It is like playing a game of dice with a die which is not only biased but whose bias changes as the game proceeds. There is no experience - no formula - on which to base the bets.

Successful systems themselves have become commodities. All kinds of manufacturing components are traded: blast furnaces, presses, robots and fabrication processes. Multi-million dollar administration software packages are sold world-wide to systemise companies' operations according to the prescriptions of the package. Production units, the heart of factories, are now bought and sold. This trade in systems requires corporations that are well supplied with both money and expertise. The expense of the systems reduces the number of possible sales so that each potential customer is the centre of intense competition. And each sale creates an "implementation project" in which the new system is shoe-horned into, or grafted onto, the customer's organisation. This project can cost the customer as much as, or considerably more than, the actual hardware and software. The game is still played by the standard rules. The competition is won by those who deliver high quality goods and services in the shortest time. Each installation project is estimated to take a certain effort and cost. These efforts are nothing less than a hypothesis that a system of type X can be installed in organisation of type Y with effort E and cost C.

The companies which compete in systems are those that are selling experiments in manufacturing science. Since they sell to all types of organisations, it is organisational science. The suppliers of these experiments - with their associated implementation experiments - need to combine high creativity with quick delivery. These software, engineering, and consulting companies require not only systems which aid administration - working out all the automatic details and costs - but ones which also provide project schedules, timetables and notification of personnel and equipment. Like all large projects, there is the ever-widening system of subcontractors who have to be scheduled. The effects of each project spread out, involving subcontracted suppliers, trainers, recruitment agencies, and in the long term, educational institutes to provide certified courses to train the trainers. Competition forces the highest and most automatic co-ordination of workers. It also requires long term perspectives to ensure the supply of trained individuals, the abundance of which provides a secure feeling for the potential customer.

The objects of the experiments in manufacturing and organisational science are themselves labyrinths of device and system logic which have evolved by incorporating best practice. The problems of consistency are ever present. The length of the evolution accumulates practices from different ages. The mix of personal styles is well known among managers. But it also occurs in the physical and logical parts of a system. Many of the techniques described as best practice do not age well. So making sure the vast system fulfils its purpose and can produce results in a fluctuating world requires intensive study of how the components can affect each other for better or for worse. The science of organisations and their systems is a science like aeronautics which is underpinned by fluid dynamics but includes materials science, engineering and systems. Organisational systems, and manufacturing particularly, are a similar tapestry, but much more general. They rely on segments of other sciences: materials science, thermodynamics, electrical technology and science, biochemistry, the mathematics of queues, schedules, and optimisation of all resources. The challenge is to keep a coherent mixture so the actual tapestry of disciplines is an identifiable subject.

If today's manufacturing systems, systems that fulfil particular requirements, are the results of a Popperian winnowing of bets, what of whole organisations?

Each organisation is an ongoing experiment. Managers do not say that this reorganisation or that restructuring is an experiment in doing things better. "Better" meaning a number of targets are met or exceeded. But that is all it is. The evolution of organisations is not trumpeted as science. The workers might object to being the guinea pigs of some bumbling manager. But vigorous, experimentally minded debate and the search for principles is the stuff of business journals and accompanies the evolution of good practice. That it does not look like "lab" science does not make it less of an experiment. The principles are another matter. It is "pre-science" or phenomenology because the principles are complicated and depend on environments which we only dimly discern. Each new successful business or new way of doing business adds to the set of likely principles relevant to a particular competitive environment.

The evolution of the modern organisation is not easily seen by the troops at their desks or in the factories. They only see the flurries of egos in the claim to some territory of "improvements". But over a fifty year perspective, the changes are clear, as are many of the principles. What was once radical, like motivating workers by treating them as responsible adults, is now normal.

Huge reorganisations such as that accomplished by Jack Welch at General Electric are grand experiments. Welch might not be thought of as an experimenter but his aim was to create a very large organisation which had to produce numbers which were much better than it had in the years preceding his ascendancy. The GE reorganisation was his big experiment which, lucky for him, paid off. In April 2000 the stock market rewarded him by giving GE the largest stock market valuation in the world. (Usually these are temporary honours and frequently have little to do with the actual worth of the company). Such reorganisations are not random castings around. The big plan is the hypothesis: an organisation of this size organised this way will let me achieve the numbers the shareholders want.

If success was guaranteed, CEOs could not ask for huge salaries. John Akers of IBM had the misfortune to see the IBM machine come unstuck faster than it could be repaired. It went from the most profitable company at the end of the 1980s, with annual profits over $US8 billion, to the record-holding loss - a total of $US23 billion - in the first half of the 1990s. Without stumbles like this there would be no "managerial mystic". Downsizing and cost cutting in technology projects is a fairly reliable way to add failures to history's list. NASA admitted as much in a television statement in March 2000, shortly after the announcement of the third consecutive failure of its Mars probes.

Information systems provide long lists of failed experiments. Any large purchase of computers and software is preceded by detailing what is wrong with the old system and why the new computer system will remedy it; the hypothesis. The system is the attempt to verify the hypothesis. About three quarters of the time this works and is advertised so in some report. The rest of the time it falsifies the hypothesis and provides the data for a report which finds excuses for the lack of success. The London Stock Exchange "Taurus" system was one such experiment which falsified its hypothesis after 250 million pounds. One of the problems here is to actually get the hypothesis right. This involves answering the right questions about the dynamics of the system. For example the integrating administrative system of the airline, motel and car rental companies which was described in Chapter 1, would have to be both consistent and reliable in all circumstances. It would have to make automatic links such as automatically pre-booking if the customer desired. Would it cope with unusual demands such as people escaping a flooded area or going to a highly popular event? Would it prevent multiple booking of motels or cars? Could it track the rental cars quickly enough to reallocate them after a clean and service? Is it expected to take room and car cleaning and servicing schedules into account?

All business change and innovation incurs some risk. It is to the constant chagrin of business managers and investors that they take the risks and the government takes its chunk of the profits free of risk. The argument against this is that, in most cases, without prior government underwriting of infrastructures, much of the business would be impossible. So while government does not incur the risk, it underpins the value. The problem of risk has two aspects: the first being that the new product or service might be rejected by the market or producing it might be much more costly that expected. The second type of risk is that which comes from the system itself. It occurs whether it is an organisation that is being changed or a valve that is being replaced. It is the familiar problem of "that little plumbing or electrical job" which encounters unexpected problems, costs, and delays. These notorious types of delays which first consume a fixed budget, then destroy all hopes of a profit and continue into the region of contractual penalties, destroy reputations, careers and companies. This is where laboratory experiments differ from organisational experiments. The difference is like the comparison between eggs and ham: the hen is involved but the pig is committed. The laboratory scientist is involved - reputations can be at stake - but failure need not bring the degree of ruin that a failed organisation can bring on its managers.

As discussed in Chapter 1, controlling risk usually means creating a wider system. One in which the experiment is monitored carefully and "contingencies" are added to the preconceived plan. These additions plan for the occurrence of problems. They are like adding lifeboats to ships and investing in coastal monitoring, the Coastguard, lighthouses, and weather stations and broadcasts. Or like adding the safety procedures in aircraft and investing in constant radar monitoring and air traffic control. They work on the hypothesis that in unfortunate circumstances, problems or disasters might be contained by "switching on" these external systems. This extends the experiment, and the hypothesis is changed to avoid failure or include disaster recovery. At worst the experiment, and usually the hypothesis, are abandoned before a failure worthy of publicising is reached. Of course the larger the hypothesis - now including the control of disaster - the more ways it can be tested to see if it fails.

This idea of a hierarchy of systems with the systems higher in the hierarchy attempting to control lower systems, is one which will become important in the next chapter. Although the "higher" systems attempt to control failure, they are more general and much weaker at controlling lower-level systems. Generally, the more events a system has to contend with, the more abstractly the classes of events are described. This is where second-order logic creeps in, with events grouped into classes according to their properties. Mechanical failures and training gaps are classes of properties of events. When events which portent disaster are described so abstractly, so then are the responses. "In the event of a mechanical failure contact the responsible officer." "Continuing operations when the weather is outside the stated range requires the permission of the officer on watch." The response of the system loses its precision. Generally, the more events a system tries to control, the weaker and less predictable that control is. The wider the system the weaker it is. Thus social systems, such as the education system, are usually "wide" and "weak" and control inputs and outputs less than specialised systems, such as pilot training, which are much more predictable. Sets of techniques to minimise and counter risks are generally weak and wide and cannot always control failing experiments.

Science loves categories. Biology, geology, astronomy, physics and chemistry put things in categories and understand these categories minutely. They categorise the interactions between the things within each category and across categories. But systems seem too diverse. They seem to resist generalisations. The vocabulary of systems science is primitive in comparison to the

111

variety of things to be described. The concepts are either too general, and end up as little more than a vocabulary, or they are too specialised, and do not apply to most systems. It is rather like grammar. Just as grammar names parts of language and insists that certain things occur together - such as plurals of pronouns occur with plural verb forms - so systems descriptions insist that certain things occur together. This is useful as an accumulation of practices, but the general principles which deliver the real insights are missing. Insights into systems can come from any direction. And some of them are "old" science. Mathematics also uses categories - it has a precise technical definition of the word. It progresses by finding the most useful definition of an idea and then studies all the different species to which the definition applies. It has abstract tools, called homologies, for ferreting out underlying structures and bringing order where variety threatens to overwhelm.

The problem has been that those with the conceptual toolkit seldom talk to those with practical problems. But this could change quickly. The dynamics of systems can look backwards for new concepts as well. For example, Le Chatelier's Principle, originally formulated for chemical thermodynamics, has numerous applications in organisations. Henri Louis Le Chatelier (1850-1936) put forward the principle that if any constraint, or force, is applied to any chemical system in equilibrium, that system tends to adjust itself to counteract or oppose the constraint. While organisations with their many interacting agents are like ecosystems with interacting species, chemical mixtures also have interacting populations of molecules. There are similarities in all large systems where numerous entities interact with each other. An event which might be interpreted as an illustration of Le Chatelier's action and reaction principle is the failure of the United States president Andrew Jackson to curtail speculation. Jackson had a loathing of debt and speculation. In 1834 he became the only leader of a major nation in modern times to ruthlessly use high tariffs and minimal public spending to eliminate national debt. He disliked the fledgling Bank of the United States which he saw as a source of speculative credit. He transferred the assets from this relatively restrained bank to State banks which had less discipline. With greater assets they printed more state bank notes for the land boom that was underway at the time. Jackson reacted by slipping through a law that required payment in silver or gold. The first result was the hoarding of gold and silver and the payment by bank notes. But as these were withdrawn, credit was called in and the United States slid into the depression of the late 1830s.

Another suggestive principle for large systems comes from ecosystems. Margelef's law is that ecosystems evolve to make the most complete use of energy flowing into the ecosystem. In practice this means if there is anything to eat some organism will evolve to make use of it. This can apply to whole economies whereby they evolve until every market - a source of business energy - is serviced profitably. The Internet as a marketing tool proceeds on the assumption that there are large but dispersed markets for a much greater variety of things than are available through local outlets such as shops and supermarkets. Internet marketing is the ecosystem equivalent of albatrosses: wide ranging creatures which feed mostly in areas others don't reach. The dream of the information system vendor is that business systems evolve to absorb all information flows, so forcing the analogy with Margelef's law.

One thing that distinguishes natural systems - considered by Le Chatelier and Margelef - from those created by humans, is purpose. Ecosystems and chemical and physical systems do not seem to have purposes - certainly no declared purpose. Human systems are created to do things. The science of systems needs to look at fitness for purpose and reliability. The first is done in a static situation - will this system, if grafted into today's techniques and methods, make things more efficient? What otherwise should count towards its success? Is it consistent with other plans? What is required to sustain it and how long might those conditions last? How well does it sit within a larger system and what potential does it have to evolve within this

Andrew Macfarlane

system? Will it be reliable? Will it demand excessive vigilance from its operators? How disciplined do its operators have to be to make it work safely?   This last question is the result of innumerable studies in safety engineering and one to which we shall return. The list of questions is absolutely practical. The most efficient way to answer them will probably not come directly. They will come from more abstract questions such as, does this system belong to the class of evolving systems, or self modelling systems? Is it a system that is determined to some precise extent by its history? We still do not know the right questions to ask. But then systems science is a neonate science.

Chapter 4: Evolving Systems. The Widening Embrace

Creating new systems is fun and easy when done in isolation. The problem comes when the complexity of the world crowds in on the elegant ideas. Each new system is a culmination of argument and experiment. Its final form is established by cutting through a forest of compromises and enforcing a social experiment with political will.

To what purpose is all the toil and bustle of the world? What is the end of avarice and ambition, of pursuit of wealth, of power, and pre-eminence?
Adam Smith, *The Theory of Moral Sentiments*

Making Systems: The Universal and the Exceptions

Change permeates life and our society. Business organisations are involved in ceaseless change. This is driven by the need for managers to be seen to add a new brick in the system edifice. Every new manager, even at the lowest levels of management, most rapidly makes his or her mark by improving the system. This restlessness within each organisation causes an overall restlessness. Managers within each organisation scan the horizon for threatening changes or ideas which they can adopt. But each deals with small changes and in the large scale no one is in control, no one person or group is responsible. Systemisation and competition need no conspiracy to spread their compass. This is not to say powerful groups do not seize the opportunities presented by the systems and technology. Large organisations, whether government, military or corporations, always have. The potential for catastrophe lies in aggregations of uncontrolled power. In consumer societies, aggregations and alignments of power are built on the general enthusiastic embrace of consumerism and convenience. The control of powerful commercial groups requires levels of thoughtful constraints and political pressure not often seen in pluralistic societies. In the meantime, as most employees work in an environment committed to surviving competition, they are constrained to work efficiently within systems and, whenever possible, to enhance systems and products.

Management is the search for effective laziness. Systems are the way to control more events and processes with less intervention and supervision. The longer the automated sequences of events and processes, the fewer the interventions, the less the supervision. This on its own would justify the effort of extending systemisation. The more processes can be standardised, the simpler the management. The outcome of this process is to build more automatic checking into the system.

Administration, management and, indeed, civic and political control, provide an impulse to develop systems with higher degrees of complexity. But complexity feeds on itself. A complex system is an experiment in making social arrangements work. It is a set of universal rules which causes a reaction by those constrained by it (so illustrating Le Chatelier's Principle). The rules are adjusted (in politics these are called amendments) with additional fine distinctions and special cases. And the more things that connect within the system, the more the changes in other parts of the network of connections need to be taken into consideration. The more things that are connected from outside the system, the more it relies on stable environments and the more control and checking it requires. Systems need to monitor as much of the network's goings on as possible. This means they have to be taught what kinds of things should be reported. This goes well beyond the first-order systems. Recall that these systems can be useful in what appear to be very complicated situations which stretch ordinary memory. In such cases a long list of alternatives has to be remembered which, historically, entailed the use of books of rules. Computers deal with them quickly and efficiently. But this is not enough when, from a myriad of events, some overall pattern has to be adduced and then something done. It is very difficult to write down the rules of children's interactions so that an alien or machine can distinguish between groups of school children in a playground involved in a noisy game, and their actually being harassed or bullied by someone. Humans can do such things easily, but working out the rules to characterise the boundary between a harmless situation and a harmful one can be difficult.

When workers, such as school teachers, can be motivated to work long hours, where is the fine line between psychologically sophisticated motivational tricks and exploitation? No doubt a rule exists, and could be written down, but the statement would be complex with very many special cases. Complexity is summarised by using broad conceptual terms with vague boundaries which then require interpretation. Enter the lawyer. By clever attention to words, events are brought inside the vague boundaries and an interpretation - and often a precedent - is established. The lawyer extends the system by clarifying its application. As was noted in Chapter two, the classification of an event to the domain of a rule is part of systemisation. The legal interpretation attempts to flesh out the complexity.

A rule or policy is formulated in terms of universal statements - statements which apply to everything in a defined class. There are no "sometimes" or "in some cases" in a universal statement. They start with, "For all...". Long hours can be spent arguing about the boundary cases before filling in the part after the "for all". The highest level of management is concerned with the creation of policies. The highest of this highest level of management, state management, the public side of which is seen in parliamentary debates, frequently involves what are called "For all / There exists " games. Here a new level of systems comes into play. It is a branch of the mathematics called Game Theory which studies the payoff involved in the creation of rules. The concept of "For all / There exists" games emerged in the area of mathematical logic as a study from the

1970s on. The motivation for its development was, typically of the subject, very abstract, but its application is very wide and is really part of the more practical Game Theory. Game Theory describes games when strategies exist to get the best payoff in systems of rules where players take turns, called "plays" or "moves", and luck plays little or no role. The theory of games was started in the 1930s and gained momentum after the war. It is not a trivial study, having attracted several Nobel Memorial Prizes in economics, and is widely used in evolutionary studies. An interesting illustration of the use of this theory is in cases where coalitions of buyers arise in auctions. This type of situation is illustrated by "The Greatest Auction Ever ". "In 1994, the US government sold off large portions of the electromagnetic spectrum to commercial users. A multiple round procedure was carefully designed by experts in the game theory of auctions to maximise both payoff to the government and the utility of the purchased wavelengths to the respective buyers. The result was highly successful, bringing more than $10 billion to the government while guaranteeing an efficient allocation of resources. By way of contrast, a similar auction in New Zealand, without such a careful game theory design, was a disaster in which the government realised only about 15 percent of the expected earnings and the wavelengths were not efficiently distributed. (In one case a New Zealand student bought a television station license for one dollar!)". This salutary story emphasises the point that our common sense, "practical" approach to social processes can seriously miss the possibilities of the situation. Indeed the design of the New Zealand auction round was flawed and although the highest bid for any of the frequencies was $NZ101 million, the total amount realised was only $NZ11 million. A similar broadcast frequency auction was held in Australia at the beginning of 2000, but with more careful design and execution, and realised over $A1.1 billion.

Games in the sense of Game Theory usually play little part of established systems. They are techniques for deciding what procedures should be part of a system. They are, if you will, part of system for the production of systems.

Rule making involves payoffs which arise when things are allocated according to the priorities which issue from agreed rules. "For all / There exist" games occur when a group who wish to introduce a policy (the "For all" group) which has to be universal, is faced with an opposition who argue against it by suggesting devastating exceptions which no one would want to ignore. This is the "There exists" group. Their "moves" are usually phrased in terms of special cases. The "For all" side wins if something of a universal rule is left. The "There exists" side wins if every proposed rule has an exception which blocks it. This type of conundrum occurs not only in legislation but also in the description of ordinary objects. How can a cat be described in terms of a computer language in a way which will distinguish it from other, similar animals? We seek a universal description of the underlying essence of the concept "cat". But exceptions undermine the essence we seek. We are used to this in politics: "Diplomatic immunity must always be observed" versus "What of the cases when the individual has committed enormous crimes against humanity" - or "The customs and ecological interactions of ethnic minorities are to be protected" versus "What of customs and interactions which will result in the extinction of a species or involve cruel and sexist rituals".

In business these games are played by those supporting existing operations. "All products or services need to pay for themselves from the start," as against "Some of our best products have been cross-funded during a start-up period". The problem also occurs when searching for threatening messages in a vast set of data. A mother, describing a school play might say over the telephone, "Johnny bombed last night. But Susan just "exploded" (on stage); she's planted the seed for something terrific (a career)". Three trigger words for national security agencies who collect each others' data and tap phone calls. The general rule, the "for all" part, is something like "highlight every conversation which has words from a given list". The "there exists" players counter with a vast list of exceptions in which the threatening word occur in an innocent context. This problem of context requires unusual levels of abstraction to describe it. The ability to describe contexts, or "domains" in computer science, makes it easier to create systems which take a context and then behave in a way appropriate to the context. The momentum of systemisation dictates that such systems will become common.

"For all / There exists" games are frequently intended to create new rules which become part of the rules or priorities by which another game is played. The object might be to grab a reasonable chunk of a political budget or argue for a set of priorities. A hierarchy of games is played out when an organisation is trying to decide under what conditions marketing is more important than new developments or projects, or whether one project is more important than another. It is played out when employment in local communities clashes with environmental concerns.

Once debates over rules and priorities have been resolved, the players get down to dividing up people, budgets, machinery or other resources by playing by the rules. The outcome of each play being the agreed allocation of a few more resources.

As noted previously, games are not part of an organisation's operational system. They are part of the process of decision making and so part of the abstract description of the dynamics of change. A device which pumps out light bulbs or car bodies doesn't have meetings. Once the settings are given, it just works. A change in the settings to produce a new type of light bulb or car body is part of the management of the system. This is what initiates meetings and snares whoever attends them. In terms of agents and messages, the boundary of the meeting, when and where it is, doesn't really exist. The subjects can be discussed informally long before people sit down in a room. Indeed the decision might not be arrived at in a formal meeting. E-mail decisions, corridor decisions, coffee pot decisions, are all commonplace. As far as the systems are concerned, a game has been played; a mechanism for deciding something has been created which can be followed. The system has additional rules. Priorities, schedules to follow, and the wherewithal to follow them have, in the fantasy of ideal worlds, become available.

Does describing meetings and decision making in terms of games add anything to the description of an organisation? Or is it a cynical ploy by this author to colonise some aspect of business consulting in order to make a lot of money? (Yes!) Although it hasn't been described in any depth, Game Theory covers all types of situations in which players can form competitive and co-operative groups to allocate resources. They can all co-operate in finding a best division of resources given some constraint. Or the can compete individually in a winner-takes-all game. If the description of meetings as games makes players more conscious of their roles - more inclined to define what they are trying to accomplish - then it is a useful insight. It is but a tiny part of a theoretical framework which can simulate much of the managerial process.

Games lie between completely free exchange "systems", such as stock or commodity markets, and deterministic systems. (The quote marks are added to distinguish the type of structure from what has been discussed as a system in this book - the rule-bound, predictable structures.) A freely exchanging but conserving "system" is one in which things are freely exchanged but the entire quantity of what is being exchanged does not change. This is rather like a tribal society in which there is no growth in goods. Accumulation of goods is pointless as it takes goods out of the trading wealth, the commonwealth, and impoverishes the rest of society. And what can you do with a stash of goods which are intended for daily use, especially when hoarding things attracts not status but odium? (This has been the economic situation for humans for most of their existence.) Games introduce a degree of accumulation. Strategies constrain future actions and a winning strategy is an algorithm which produces a winning situation whatever the competitors throw at it. Thus a game with a winning strategy, unlike the systems just described, is one in which each input - a new move in the game - has a predictable response. A winning strategy is an algorithm and makes a game a system in the sense used in this book.

The famous management consultant Peter Drucker described management as a humanities subject. The study of management is a liberal art. And this should be so at the highest levels of management where policy is formulated. It is here that intentions and purposes are formulated. They are put into statements that can be implemented through policies. These policies are then the basis for strategic thinking. This is the domain, if not of the philosopher king, then at least the thinker-CEO. This is the creative act of guiding the corporation, reforming it and turning it into a lean, mean, shareholders' dream machine, or turning a government department into an effective, thrifty, humane arm of national administration.

Perhaps.

There are a number of constraints. The first is accountability. It comes in several ways. The highest level of management is scrutinised for its performance. Unless you own the firm, you are an employee, and not necessarily a tenured one at that. Even if you own it you are vulnerable to takeovers. Many top managers are on contract with a set of performance measures to meet. They are just the top level operators of the machine.

An aspect of accountability is that whatever the ends, the means have to be acceptable. They not only have to be legal but also have to be understandable to the audience of company watchers, analysts and investors. Changing the machine which you, the CEO, command, requires that you do it with approved techniques. The existence of MBAs, postgraduate business degrees of all sorts and consultants (the MBA career choice) suggests there are techniques. After all you cannot have a qualification without some technique. MBAs wouldn't be so popular if they were only travellers' stories and fashionable parables from the corporate story book. The observers of the top management will watch for action. The re-organisation of marketing or production. They will watch for the forging of alliances, inter-organisational systems, takeovers and makeovers, the extension and re-organisation of systems, the bold experiments (when approved) or risky gambles (when not). The actions will be discussed as applications of technique or criticised as risky departures from established principles.

The top level of management is the one in the public eye but, as already noted, every new manager is under pressure to change systems. Constant change and reorganisation are epidemics which suggest that managers do not clearly see what structures they want. The constant experiments in structure have less to do with making systems more effective than to do with short terms gains and making political statements.

Systems themselves force constraints on a manager's actions. The larger the system the longer the commitment to it. You don't mess with an administrative system which cost scores of millions of dollars and was approved by the board of directors. If you think it is an albatross around your neck, live with it and just try to make it work better. Much to the annoyance of many an ambitious manager, these systems become fortresses from which others defend their reputation and territory. But personalities aside, a system of say 10, 20 or 50 million dollars, which took three, four, five or more years to wedge into all the nooks and crannies of the organisation, cannot be dismissed easily. It has to be seriously and publicly defective before it becomes a moveable object. These systems form a political constraint on change. They represent huge inertia and resistance to change. Unlike engineering or production systems, which can be ten or fifty times the investment, administrative systems do not represent a defining aspect of the organisation. The factory plant defines immediately what can be done. Investment in new plant can be a do or die decision. A twenty million dollar generic software package represents no such thing.

The constraints do not finish at the organisation's doors. As the Internet and electronic commerce systems become widespread, any change of administrative systems will face swarms of technical arguments on what additional things have to happen to remain compatible with everyone else's systems. As organisations dissolve the boundaries between each other, and their systems become interconnected, changing them will be like changing the dance steps on a crowded dance floor. It is only possible if everyone does it together or you have to work your way to the outside before you can have such freedom. The flamboyant entrepreneur and corporate creator, dancing his or her own steps during eras of transition, eventually get squeezed off the dance floor by the adherents of approved technique.

Context - Rationality Beyond the System.

People are rational or crazy. They can explain their actions according to some ethical strictures or a set of understandable principles. We expect every action which requires some effort to have some payoff. This is our social context of rationality. Even if it is only for the good of the soul, or to preserve or ameliorate something, there is some payoff. The reasons might appeal to ill-defined or apparently contradictory principles, but they will rest on increasing wealth, power, status (which might be ascetic and spiritual) or security. People whose actions cannot be explained by some such principles are deemed "crazy". They are taken to be *irrational.* Our social expectations of rationality are violated. In business and politics, where it is normal to expect an account of why an action was chosen or should be chosen, the principles are explicit; no one is beyond accountability and the games have materialistic payoffs.

Rationality lies in games of wealth, power and influence. It restricts the games we play to those which have payoffs in its terms. The collection of all the allowable games is the logical expression of this rationality. These, in turn, are what gives reasons for, and envelopes, all the types of systems which have been the object of our discussion. This rationality is the broadest and haziest context with which business and political leaders live. It gives the reasons for creating and changing systems. Systems are now the instruments of wealth, power and influence. Conquest and theft in the name of your nation has been written out of rationality. Heroism and courage to save or preserve something has not. Our commercial rationality has the germ of progress.

The creation of systems has been described as a process of hypothesis and experiment. It is also a play in the game of market share and return on shareholders' investments. This is no contradiction. A chess player makes a move based on the hypothetical evolution of the chess game. The opponent makes a countering move intent on falsifying the hypothesis. It is still a game with each move having a payoff. Similarly with the creation of systems. The system is a managerial play which attempts to gain some stable advantage - at least for the next couple of moves.

For most people, the context in which they work is reasonably constrained. Most of us live in the comfort of a "sub-game" where the context is defined by some larger system containing the system which defines day to day work. If you have an understanding of the larger system then you know your options. Working out the best option is easy. It is a question of listing things which take you closer to your goal; listing moves with the best payoff. This is what we usually do when the number of possible actions or moves is small. This is the "plan A and in case this fails plan B or at worst plan C" type of discussion. But too often the number of choices can involve not only a large number of factors but ones which all have a bearing on one another. And one way to hold all these together is to formulate policies which then give some order and some sense of priority.

High level management creates structures with policies, whether via a debate or a game, in order to have them realised in the day to day form of operational systems. Although each system is an experiment, rationality demands that it is not plucked out of the air. And although there is a hypothesis that it will fulfil, at least, the wishes of management, if not the wider set of stakeholders in the organisation, what guides the process? We have mentioned that high level management has its priests - the professors of business schools and the top business consultants who advise on markets, finance and motivation - but in terms of systems, what makes the good bet? In an age of systems what are the systems which rank systems as being good or bad?

This is the idea of optimisation, or the best fit, the best practice, the better way. It has its intellectual ancestry in the systems analysis and operations research which was developed during the Second World War and the Cold War. Many of its techniques preceded those two events by centuries but the two contests ushered them out of learned journals onto the public stage. The goal of optimising is to find the best or worst solution among a set of possible solutions *according to some measurement.* An example of this is the travelling salesman who has to travel to a number of cities in a large area and no city is to be visited more than once. The measure to be minimised in this case is the total length of the journey. The solution to the problem relies on this measurement. The actual salesman might have other measurements in mind such as scenery or social contacts in towns along the way. In this example the set of all possible routes makes up the "state space." The state space is all the possible ways something might be solved. There are techniques which can work efficiently through the state space to find the best route. When a couple moves into a house, each imagines an arrangement of furniture and household items which fits their personality. Each imagines an "obvious" arrangement. It doesn't take long before the details of one person's "obvious" are not obvious to the other. All the possible obvious, non-obvious and downright crazy arrangements of furniture constitute the state space. Each partner has an idea of convenience which becomes the measure to optimise. If the measures of convenience clash, the chances of a harmonious optimal arrangement are not good. A little bit of game playing might be used to establish which measure of convenience is to be used. If the arrangement of furniture in a household has many tricky options, imagine what it is like rearranging an organisation.

Optimisation is a deeply held brick of rationality. Politics sets out to create policies by which certain ways of doing things are taken to be fairer, more equitable, to provide a better business environment or natural environment, to free up investment money or to make us more secure (which is surely a string of failed experiments). Business creates systems to be more efficient, reach more customers, obtain more investment money and so on. As long as the "more" is one of the accepted measures of rationality, no further justification is needed. The means might be subject to debate but the end is not. If the arguing parties agree on the measure of improvement then the demonstration that the new way is optimum clinches the debate.

There are many types of optimisation problems which have been solved by operations research. These have been incorporated into the design of manufacturing and distribution systems throughout the world. They are part of the tools of the trade of operations management and engineering. But at this stage they have hardly dealt with whole systems.

There is in fact no clear barrier to optimising systems as a whole. The same techniques, requiring one or more measurements to be minimised or maximised and a concept of the finite state space are all that is required. This last idea is that the number of options or possibilities is constrained by some additional "on or off" consideration such as whether a possible solution is politically unacceptable. Crazy possibilities need not apply. Such rational or practical constraints include the idea that the system should not be too difficult for humans to operate or too expensive to attain. But these are state space problems.

127

It turns out that some ordinary thinking can get us to the best option which operations research will confirm. Not always, but often. We do not know how true this is of systems or organisations. The map of conceivable options is the starting point, and, as pointed out in the previous section, one generation's "natural way of doing things" was crazy and radical for their parents' or grandparents' generation. As with the not-so-obvious nature of the arrangements of household furniture, the conceivable systems and organisations are not necessarily obvious. Especially when such things are poorly described or complex to start with. And the problem can be compounded by asking that we optimise something which is difficult to measure but very desirable, such as *capability*.

The optimisation of systems or organisations does require something more than the previous techniques of optimisation, if only because of the way they involve people. The first thing is to add to the set of all the organisations we wish to consider, the ways in which they could evolve into one another. It is easier to have a vision of where you want to be than to map out a way to get there. It is often a sign of sloppy political thinking if a party paints a utopian vision but cannot supply the road map to get there. The evolution of organisational systems, while not necessarily obvious, usually involves simplifying sequences of tasks and then combining them together, and the contrary trend of specialising and splitting them. The first occurs when several previously separate tasks are shown to be variations on a theme which can be automated. The contrary trend occurs when things become too complicated or refinements in processes are made which require separate jobs. Both of these types of changes can be inspired by changes in technology, such as devices which can recognise items or move material more rapidly and accurately to where it is to be used. The astute manager - with an instinct for systems - will be on the lookout for useful technologies, if not actually supporting the development of them.

However the evolution occurs, there is some financial, personal or political cost. No change comes with nothing but accolades. We shall call this penalty "resistance" so that it is not specifically identified with a financial cost - which it might involve. It is a more general concept of Le Chatelier's tendency of systems to resist constraints as a forced movement is a constraint in a particular direction. It can be visualised by thinking of the organisational landscape as being soft and saggy. Staying too long in the same place - not adapting - means that one starts to sink into the spot you are occupying. Eventually, climbing out and moving on takes a great deal of effort. That is resistance. It hit IBM in the early 1990s when changing computer fashions destroyed the market it had so lovingly made and had structured itself to nurture. After suffering record losses it climbed out of the mire it was in. A huge, painful effort was needed.

What we have described is what is called a *category* of organisations. It gives the ways in which one organisation might transform into another which is the start of describing the possible evolutions of a system. And it gives the resistance. Suppose we have decided on a measure of desirable structure which an organisation ought to have. If we can identify the abstract description of our own organisation, something which can be quite difficult - as even managers know little of their organisation away from their territory - then we can chart the optimal course to a more desirable one. We might have to choose a path of least resistance but again that is not hard to do. What is hard is actually using the descriptions. They are inevitably very abstract. It can be done but it is a very unfamiliar activity - at least at this stage of systemisation.

For those that think this is all very abstract, all very removed from practical concerns, consider the problem of moving from a hardware supermarket to an Internet based distributor of hardware, which guarantees next day delivery to any corner of the nation. Resistance comes from competing against your own shop-based business, possibly out-competing this aspect of the business, reducing it to a minor part of the whole business. At the same time the historically profitable business is being reduced, money is borrowed to invest in distribution links. What is the path of least resistance for a given time frame? How optimal is a series of intermediate steps in an environment of world-wide commodification of hardware goods? Is some type of customer advice, online or otherwise, part of the capability of the new structure?

Having solved that one, move on to the evolution of universities as they become providers to huge news and entertainment corporations. The resistance is enormous. Lecturers become content providers working in teams with animators, drama producers and graphic artists. Tutors toil to keep up with their quota of Internet "clients". The properties can be capitalised as research parks as most students are enrolled online from any part of the world. Researchers have to get used to funding their work from consulting fees or through departments which allocate shares in the universities' royalties for intellectual property. At what step in this evolution does University A decide that it is going for broke to be the number one language and international business franchise, and dump its science faculty? Where is the path of least resistance for University B which is going to be the top class provider of science and medicine and dispense with its humanities faculties? Does the news corporation pay the entertaining lecturer more or less than professors who are leaden before a camera? How does one market research capability, and who pays for it, when the university is part of a world-wide franchise mega-educational delivery group? How does one replenish the research teams with new talent when young researchers see that intellectual life does exist outside corporate universities and offers enormous opportunities? Why should the taxpayers of any one nation subsidise it? Finally what is being optimised and for whom?

The eminent social scientist Kurt Lewin observed that, "There is nothing so practical as a good theory", for a good theory organises that part of the world and facilitates decision making. However, more mundane than the history making examples, are the problems of everyday high-level managers searching for better ways to improve their devices. All this restructuring, business process re-engineering, business process improvement, is an attempt to extend the system concepts to more and more of the organisation. Quality management systems (see later) are such an attempt, so is e-commerce - it cuts out structure between the buyers and the suppliers. All the ideas, seemingly so radical initially, are seen as being relatively obvious in retrospect. This is because we haven't fully discerned the principles. But generally the day to day restructuring involves further specialisation of skills or, conversely, simplifying or merging of previously separate skills. These can be seen as moves in a game which has efficiency as its goal. More people are added as products and processes become more complicated. Or, conversely, new products or processes allow the merging of separate skills to allow a leaner team to be produced. This usually results in redundancies. In both specialisation and simplification, or merging, there is some resistance. It might be from the financial cost of change, it might be the cost of retraining or the redundancy payoffs. Resistance relates to the social and system effort required to accomplish the change. In order to reduce it, it often pays to bear the cost of specialists in helping people through the change.

This is change management. Strategy identifies a desired structure which has a payoff: it will make the company more robust against competition or it will reduce running costs or simplify the systems and make them more flexible and so on. There is always a penalty to the effort of becoming the strategist's desired structure. There is a resistance to movement - financial costs, political costs, the problem of re-training the workforce or retrenching them.

These ideas are commonplace, from dreamers to hard nosed managers. But they are not in the realm of standard concepts which can be manipulated, communicated and calculated. Concepts in systems evolve, refined by failures and strengthened by successes to become standardised components of our intellectual machinery. The concept of the state space was standardised by dressing it in the respectable clothes of the mathematics of vector spaces and manifolds. The least resistance evolution of organisations can be dressed up, quite respectably, in the mathematics of category theory and model theory. The ideas are in place but practical application will deepen and extend these abstractions. Which in turn will formalise the rationality of the systems of society.

Countering The Golden Rule

The Golden Rule: *He who has the gold makes the rules*. (Anon)

As jobs become more specialised they become more vulnerable to changes in the economy. Farmers and primary manufactures at the foundation of the economy provide much of the real wealth that filters up to pay for tourism, health, legal and educational activities. When the upward drift of money ebbs, these upper reaches of the economy can develop dry regions which become the focus of retrenchment and unemployment. That many of these people are articulate and media-wise gives an exaggeration of their plight. This is the politics of the instability of the modern economy.

Systems work best when they are in stable environments which ensure that inputs are standard. This happens when they are encased in, and protected by, a larger system. An otherwise efficient agribusiness, intent on supplying abundant supermarkets, can be degraded by poor roads, uncertain supplies of fertiliser and an inadequate veterinary service. The supermarket requires a steady supply of electricity, good plumbing and, more recently, electronic links to bank data for EFTPOS operations. If any of these supplies and structures are unreliable, profits are lost and prices go up. Our society is highly ordered and depends on infrastructure which costs hundreds of billions of dollars to maintain. This infrastructure includes the supply of water with levels of bacteria far below what humans have drunk before, roads that are smooth enough to drive for thousands of kilometres, at 100 kilometres an hour, without a significant bump, and an international financial system by which we can effortlessly buy things around the world from the comfort of our own homes. This astonishing level of convenience, stability and order depends on the management of an underlying stable infrastructure, all of which started in the nineteenth century. This included water, roads, power and the financial system. It is of course a predatory economy which lives off the cheap labour of people in developing countries, most of whom haven't a hope of buying what they make for others. This now common trick is promulgated by sleazy, lazy managers. It exports the pre-Ford era of industrialism, the sweat shops of the Victorian era of oppression, to Mexico, Indonesia, the Philippines and wherever desperation for work creates cheap compliant labour.

Where complex systems have developed so has the economy. But this development has depended on some social conditions to foster it. The evolution of systems of systems can only exist in a highly structured society. This does not mean a totalitarian society. Even the anarchic, "trash the system" spirit of pop music relies on a vast array of relatively easily available technology to create, publicise and broadcast its creations. This technology is invented at the end of a long sequence of other inventions and social innovations. Inventions which have occurred and been adopted in all sorts of conditions.

The extent that a society has structure, which means its laws, its infrastructure of roads, communication and the supplies of the necessities of life, is a measure of its stable systems. This stability and structure allows the free market mechanisms which have so mesmerised economists. The theoretical free market, as the antithesis of system and order, only exists briefly in the unusual circumstances in which traders freely exchange information about prices and no one is in a position to maintain starting advantages of prior size, wealth and connections. These advantages have always conferred "non-equilibrium" conditions where blocs of traders can dominate the market. No society has ever satisfied the circumstances of a theoretical free market.

Economics, as the study of the conditions of wealth, has concentrated on the intellectual legacy of Adam Smith, who hypothesised that the primary precursor of wealth was simply markets without fetters. This is not the experience of recent and ancient history. Managerial despots have often left their countries wealthier for organising systems which allowed trade to be conducted more easily. This has been a frequent story in Asia. The people at the top might be corrupt but at least the gangsters at the top suppress gangsters lower down. They then push their societies into rapid levels of systemisation. This allows the rest of the population to invest in and create a stable and homogeneous environment.

Systemisation is more the key to wealth creation than unfettered exchange. After all, a third world village might be a model free market, but it doesn't escape the poverty cycle. While it is true that the United States has been *the* wealthy country of the twentieth century, the adoption of its methods of mass manufacturing in both free and non-free circumstances has been the crucial creator of wealth. Even the antithesis of free exchange, communism, has had a press that in economic terms is unfair. Soviet Russia grew from a non-entity in terms of world power in 1917 to the second most powerful nation in 33 years. This in spite of the appalling damage done to it by two of history's most evil people, Stalin and Hitler. At the cost of 20 million dead in the Second World War, and with no postwar transfer of American wealth - as with Germany and, to a lesser extent Japan - Russia not only survived but emerged as the number two superpower. But the stress of the arms race of the Cold War kept standards of living low. Stalin's heritage of a generation of party hacks masquerading as managers was not up to the job of balancing the demands of maintaining a huge war system at the same time as producing consumer items in abundance. The collapse of communism is regarded as a vindication of capitalism rather than a lesson of totalitarianism. In fact had the communist system not been devastated, and not had to compete against a nation whose industrial base was not bombed and invaded, it might have a better reputation. Indeed China, after flailing around in a revolutionary fervour, salted its communism with some market forces and produced a growth rate of eight to ten percent for two decades. This is not to defend its totalitarianism, but to point out that as a social structure for deploying the fruits of systemisation to create wealth, its reputation for disaster is not deserved. And indeed it is the deployment of social systems which seems to be much closer to the key of wealth creation than merely relying on free markets.

An example of a government initiating outright systemisation is Singapore, which under Lee Kwan Yew was not a shining light of democracy. Singapore's TradeNet is a network which links the management and operations activities of the world's largest port. The Singapore government initiated and spent approximately $50 million in the 1970s on the system which linked trade-agents - freight forwarders, shipping companies, banks and insurance companies - with the relevant government agencies - mainly customs and immigration. It used to take a ship up to four days to clear the port; now it takes as little as 10 minutes. This dramatic reduction, in time, made Singapore the port of choice in South East Asia. The success of Lee Kwan Yew's rule as CEO of corporate Singapore can be seen from the change between 1965 and 1988. Population grew from 1.89 million to 2.65 million, gross domestic product (GDP) per person went from 1,692 dollars to 15,999. Students at school rose from 15,000 to 77,000 while the labour force grew from 38 percent of the population — 723,000 to 1,282,000 or 48 percent of the population. The amount of sea cargo handled went from 15.1 million tons to 142 million, with foreign reserves climbing from 14.2 million to 3.34 billion dollars.

This is just one of many initiatives taken to produce large systems which become strategic aspects of a nation's economy. In Korea and Japan, the government favoured corporations such as the group of Korean *cheobol* corporations and the Japanese pre-war *Ziabatsu* and postwar *kigyo shudan* or business groups. Unfortunately this led to the undercapitalised loan-dependent corporations which undermined the banking structures of Japan and Korea in the late 1990s. This distortion of free markets - pejoratively referred to as "crony capitalism - favoured concentrations of modern systematised producers. When systemisation genuinely covers the banking sectors of those countries, their strong production systems will be intact.

The system of trade that prevailed from the end of the Second World War until about 1970 was based on fixed exchange rates - a standardisation of currency. Since 1971 a progressive de-standardisation has occurred. Currencies float freely and can sink or soar depending on perceptions, hunches and, over the long term, those hopefully rational principles called "fundamentals" - the set of economic measurements which theoretically give a sense of value to a currency. It is curious in an age where systems and standardisation have shown how useful they can be, that the media of exchange are free of standards. This is resolved in a new level of systems which is the start of another story.

Living in a world of floating currencies has not proved to be a burden for those well armed with systems. The systems are the lifeboats in stormy seas. As currencies have abandoned standard ratios, the international banking and currency networks have provided close "real-time" or minute by minute tracking of these ratios. The art of hedging has become more and more analytical. It is easy to buy currency at given rates of exchange in the future. There are financial "instruments", such as buying forward and options which allow all this to be worked out. The valuing of options and derivatives has spurred the development of sophisticated financial mathematics and has contributed to a growing system to "manage" - basically minimise - the losses which can happen when storms hit the financial sea.

Gambling with future exchange rates and commodities prices is a respectable occupation. Much of the excitement of the Internet arises from the creation of exchanges of various types - and from there to a widening array of commodity markets. Why not lifestyles: future education prices, health insurance or insurance against skills becoming obsolete? All these things would be regarded as an opening up of the market. This could be driven by the argument that management by the actual users of the financial services will not only be more efficient (by what measure?), but will free money for governments to provide pensions for an ageing population. To make some success out of this experiment, these social structures could rely on networks of databases supplying information in much the same way as commodity data are made available. One can speculate that future governments will aim to reduce their responsibilities for certain standards and services if it can be shown that they can be managed by some outside agencies. The agencies need not be centralised but groups who accrue some advantage by creating some public good or social service via a networked system which acts as the glue for the participants. The entire "government services" sector could be private organisations, attuned to, and specialists in, their slice of the service pie, and with access to their part of the greater government databases. The obvious drawback is that private data would accumulate in the databases of those motivated by profit. Privacy would erode as both government auditors and private market researchers sought to lay hands on the data.

Even if governments transfer the funding and provisioning of future services to networks of corporations and financiers, they are still the caretakers of the state justice systems. As long as the nation-state exists, no unelected corporation can substitute for an honest government's role as an impartial body to adjudicate between organisations beholden to their major shareholders. Nor can any profit-oriented group do more than advise the judiciary, and only then on request. And no private corporation can provide an environment of stability. If corporations are the subsystems, then an overall controlling - or at least adjudicating and harmonising - body is required. This will not be an easy task - the games of allocation will become more complex. The pressure group activities of those wanting to adjust the system or make room for those who promise to do things better will be acute. The funding of pensions, health and, to a lesser extent, education, will require sustained growth for some time. The age distribution of the population will see to that.

Legislation has to be understood, open, logical and have a defined set of priorities. It has to be part of, or create, a system. Administering governmental social systems is the province of the civil service - the (often wrongly) deprecated bureaucracy. It also should strive to be *logically comprehensive*: to apply universally and, when disputes arise, to be able to resolve them without ambiguity. The fact that the West has a legal system based on precedence means that the legal system is never finished and is enhanced on a case by case basis. In system terms it is not complete. The For all / There exist games have not been resolved. And this is how society has muddled along. Edmund Burke (1729 - 1797) saw this as a healthy thing. He saw the attempt to create a society based on a logical system as being incapable of dealing with the richness of life. This insight remains unshaken. In practical terms any system which includes the wisdom and hairsplitting of case law, needs experts to navigate it - experts such as lawyers.

Government creates systems by legislation. These include the rules of exchange by which the economy becomes a system. Civil servants advise and lawyers revise. The growth of the legal professions in countries whose legislative traditions come from Europe relies on overloaded legislative systems. Where the language of law is fuzzy, with arguable interpretations of detail, lawyers proliferate. They prosper when inconsistency arises from shades of interpretation. Above all they prosper with legal complexity. Bartolo in the Marriage of Figaro stands as the archetypical lawyer as he sings, "I'll equivocate and paraphrase and trap him in a legal maze". But need this be the fate of modern society? Large corporations do not have many lawyers to interpret their rules, which admittedly are simpler but also written more directly. This suggests that there is some possibility to do better. Is it always so that social systems - the rules of getting on in society - should be so complex that the details need be overwhelming? Is this an intractable aspect of life or is it possible that powerful systems might at least aid the good and fair running of society? Such systems would not supplant the creative process of decision making but would help validate it against desired outcomes. With some exceptions, engineers do not release systems until they are tested. Legislation is debated and the drafting is done by legal professionals. Then it is "released" by an act of parliament to be tested on the populace at large. Like the exception in engineering - civil engineering, with its dams, bridges and roads - the legislators have to test their social systems "in place". There is no conceptual barrier to producing systems which simulate various types of behaviour which could be run through legislative systems to check for consistency or fairness. Fair allocation is a subject well worked over by Game theorists. Systems using these principles could use the results of many varieties of games to check the deviation from fairness. These systems could only ever be an aid to the auditing of legislation from agreed principles of fairness, concepts of property and expectations of security. They would require much initial research and development, but then would be part of the running costs of a democracy.

Many would shrink in horror to think that a system might interpret the rules by which we play. Systems are thought of as cold, inhumane, and brutally impartial. (Justice is supposed to be blind but compassionate.) It is not systems per se that lead to man's inhumanity to man. Indeed the very act of elaborating the rules and checking for inconsistencies is probably of some benefit. It *is* possible to write into a system classes of people to be treated differently. But at least it is explicit. It makes gross hypocrisy explicit. The famous national hypocrisy of slave-owning Christians or even the slave-owning creators of the Constitution of the USA would be sharpened.

The success of many constitutional arrangements and of case law should not blind us to the magnitude of the task of "getting it right" in a world of global competition, environmental and trade restraints, and legal, environmental and financial obligations. To get it right is a huge burden on legislators. If humans are so wise in their social and political arrangements, why do they change all the time? When do problems of complexity and scale force us to look for additional conceptual help? When cartels of large corporations lobby, with all the sophistry which money can buy, what do governments, with their smaller budgets, do in response? Finally, even with the most sophisticated systems tools, who is in a position to audit the tools?

The modern economic system presents itself to its managers as a vast array of competitors, all finely tuned to each other. Where it is very different to a machine is that it has evolved, and evolution is a game of destructive testing. Failure and death are closely linked. Managing a modern economy is managing a system of hundreds of subtle feedbacks. Democracies are so finely tuned that a change in interest rates of ¼ to ½ of a percent can strip billions off the apparent wealth of a country in a stock market hiccup. Currency shifts are reported nightly on prime-time television, with the changes of more than one part per thousand being serious news. No operators of factory systems need watch temperature gauges for such fine changes. Changes in the altitude of a jet airliner of one percent - 300 feet at 30,000 feet - can be caused by air pressure changes. The impression is given that the modern economy is a precision machine with which the politicians have to play to adjust social policies. When it is going well - when injustices are not glaring - only a fool would tamper with such a machine. As the global economy becomes more complex and more tightly connected, a conservative attitude will harden. Only if there is a lot to lose by not playing the economic machine will politicians act.

Chapter 5: Systemisation for Safety?

Failure modes haunt every advance in technology. And every new technology creates a new social experiment. Drugs, however well tested, only show their nasty side-effects after sufficiently many people have taken them. Problems and outright disasters produce new science. Exploding boilers and cannons pushed the development of metallurgy. So did metal fatigue in the Comet airliner and neutron embrittlement in Nuclear Reactors. Technology is imbedded in systems of correct procedures for use and maintenance. These systems require that their adherents accept their disciplines or face awful consequences.

> For want of a nail, a shoe was lost. For want of shoe, a horse was lost. For want of a horse, a message was lost. For want of a message, a battle was lost. For want of a battle, a war was lost. For want of a war, a kingdom was lost. And all for the want of a nail.

Creating High Quality Complexity

The theme of this book is that systems have been with us for a long time although we have been creating them at an increasing pace. Today, systems are the way we solve problems. The great systems of abundance - the triumphs of engineers from the start of the industrial age - have all been derived from the mechanical, the physical and the chemical. They have been able to be visualised. Or, at least, they rely on scientific principles which can be visualised. An oil refinery can be explained with a small knowledge of chemistry and a good guided tour. A steel refinery or an assembly plant is similarly easily understood. This began to change with the advent of the computer. Control passed from clever physical devices which exploited pressure, magnetic force, gravity, springs and levers to control other devices, to digital devices which encoded the logic of physical controls. This was the start of the logical system. It has delivered an age of invisible mechanisms. One can look at a car and an aeroplane and, possibly a factory, and get a feeling of one being more complicated than another. Is a word processing program more or less complicated than a large payroll system or a system to control a power station? Comparisons of complexity depend on counts of components and the number of connection pieces between components. You also want to count the number of ways one thing can change when affected by other things. This is not an easy exercise for even familiar items such as refrigerators. It is a specialised abstract task when applied to a computer program.

Understanding complex systems which use the flow of fluids or chemical reactions can present problems. Aircraft use the flow of a fluid, air, for flight - simple? How many can explain the principles of lift or the drag of vortices from the wings or the creation and effects of the shock wave as the aircraft hits the sound barrier? How many ways can the smelting and casting of alloys go wrong? How does one control the manufacture of dangerous chemicals such as ammonia, strong disinfectants or herbicides when reactions must be finely adjusted for chemical reaction rate, heat and pressure? The scientific principles are comparatively easy to explain but the scale on which they have to be mastered is difficult. Mass production means large quantities of raw material which will inevitably vary in quality and mix unevenly. A ballpoint pen factory might not seem very glamorous, but the tolerances of the ball in its socket and the viscosity of the ink make the whole process one of extreme precision. The control of the process might require multiple checks and tests but the idea behind each step is clear and easily explained.

In contrast, understanding the details of even a simple program can take a specialist days. Ordinary administrative systems take months or years to understand fully. Mathematics makes it clear that understanding abstractions is difficult. Reading a university mathematics book is a daunting prospect. But the full documentation of moderate administrative systems can take a roomful of documents devoted to systems requirements, design diagrams and program code of the same difficulty as the mathematics text. It can be so extensive that no one knows it all and experts take years to accrue familiarity.

An illustration of this is the Y2K problem. Solving this problem consumed something in the order of $100 billion dollars in the United states alone. The fact that this figure is a guess, and bitter debate as to whether it was all a "con" indicate how poor is our social level of understanding of systems. Certainly the problem is simple to describe. Calculating times between two dates is simple and ubiquitous in administrative and control systems. These calculations fail if the years are represented as two digits and the dates occur in different centuries. Programmers had known about this for decades before the event. But programming languages were written in a way that made the two digit year standard and easy to use, and four digits required extra work. It was the wrong standard but no one knew the effects of any prescribed amount of effort. What was too much and what was too little? How long would it take? How much testing did one do? Without some concerted effort, a blizzard of irritations, large and small, would have issued throughout the year 2000. The media were aggrieved at the lack of disasters. Without a "control" country to run the "let's ignore the whole thing" side of the experiment, we shall never know. As with economics, the experiment would involve sacrificial or "guinea pig" communities. (Indonesia was reported to have decided to ignore the whole thing but in the overall disorganised state of the country at the time such administrative problems would have been a minor consideration).

Computer systems present the problem of vast invisible (though not unimaginable) sets of connections. The fact that these are abstract means that high-level descriptions of intent can themselves suffer from too much detail. There is no clear "scale model" of a computer system as there can be scale models of ships and boats and planes, not to mention factories, oil drilling platforms and nuclear power stations. These scale models of physical systems can be used to check the consistency of the design and are useful to teach people how the plant works. Certainly there are useful diagrams which can be made of computer systems but they can have scores, hundreds or thousands of "objects" representing stores of data or functions, and hundreds of links between these items. There is no "see it at a glance" understanding. And this of course is the most easily understood "top-level" description of any worth.

Administrative computer systems can be a challenge in their complexity, but they generally don't threaten anyone when something goes wrong. The combination of physical and digital systems is another story. The complexity of each is compounded by the connections between the physical and the logical. The physical, frequently being dangerous if not controlled, usually demands levels of discipline not associated with administrative systems. As we shall see, oil platforms, factories and aircraft kill people when not handled carefully, when work practices are violated even in trivial ways. Add to that the uncertainty of unreliable gauges, inexperience with unfamiliar states of the device, and unclear computer representations and, as will be discussed below, system nightmares are realised.

The systematised world is increasingly one which is represented on a computer screen. Knowledge of the real effects and events is mediated by systems which only the designers and programmers understand. Sets of interacting numbers are hard to control however they are obtained. Factory and device operators have to gain a level of experience of a filtered representation of the physical world. The direct experience of the physical is no longer efficient and cost effective. What we know depends on how much it costs to know.

The operation of large systems is so complex that the designers, who must know the devices better than anyone, are

145

forced to put layers of automated controls which limit the operator's options. This is what is behind the familiar "invalid values" messages in so many administrative systems. It adds a frustrating layer of rules which have to be learned before the system is useful. These layers are put there at the behest of customers but the flux of staff in large companies, together with extensive and obscure documentation, means the social knowledge of the evolution and reasons for the system is lost. This social knowledge is the way we approach complexity. It is much more human to ask someone who knows "the system" than to read manuals.

But the operation of the systems is simple in comparison to their creation.

Only a few generations ago the actual manufacture of cars, farm machinery and implements, aeroplanes, domestic appliances and toys was the daily lot of armies of workers. This is now substantially automated. The remaining labour intensive aspects of manufacture have raise their intellectual stakes. It is the custom built creation of systems which involves people who now work at manufacturing, but they are far from assembly lines. The engineers, designers, decorators, computer programmers, financiers, project managers, quality managers, auditors and testers now build the systems, devices and factories which have replaced the workers. They work at a higher intellectual level because they have not yet standardised and systematised the higher skills needed to solve new design problems, how to use new materials or old materials in new ways, even how to test new devices, systems and factories. This standardisation and systemisation might never happen. If one level of components becomes so well understood as to be standardised, it becomes part of a bigger more integrated device which is not at all standard. Thus creative and problem solving aspects of this activity might forever be an area of skill out of the range of systems. However the social process of creating systems is well on the way to becoming systematised.

All systems present the same problems. The first is how to describe them - how do you describe something you haven't seen. This is the problem of linking purpose and specification. Once the description is agreed, how are the teams of specialists, and companies which supply parts, to be co-ordinated? How are the components and assemblies to be tested? One cannot simply hand these systems over to the customer and walk away from them; so what is involved in the implementation and the development of appropriate operating practices?

Even the problem of description is not easy. Any new system has to have a purpose which is not found within the system itself.

In order to bring these ideas into sharper focus, consider a device such as a rocket. It has a specification which is to lift a minimum payload to a certain height within a certain time after starting. This is the specification - the big scale of things. The rocket itself uses solid fuel or liquid fuel. If it is liquid fuelled, it might be composed of two or more fuel tanks, a rocket engine and fuel pump and some controlling electronics.

The specification of the rocket in terms of its purpose - the requirement to transport the payload as required - is one level of description. At this level one can make comments about the behaviour of the whole device such as, it exploded on takeoff, it overshot its mark, it takes forever to fuel and so on. The description of the subsystems which arise from its specification uses a different "vocabulary". A certain thrust is required; this in turn requires a certain performance from the engine, a certain amount of fuel per second and so pumps and fuel tanks of certain size. The components are at a different level. The strength of the metal which makes up the rocket engine, is described with different specifications to the specifications of the engine itself. Again, they are at a different level.

The rocket is a composition of pieces, each with its own set of specifications, which, in turn, implies that the device will satisfy the overall specifications. What *emerges* out of the composition is the overall capability of the rocket.

This is all "common sense." The overall characteristics of the rocket can have puzzles. What level of control is required to keep such a thing stable while its centre of gravity changes as the fuel is burned? How does one work out controls which will not over-compensate and produce a gyrating trajectory which is impossible to control? A test of understanding of our technology is to ask if you could reinvent it. How many domestic appliances could the average householder reinvent? In almost all cases the reinvented device is simpler than what it takes to get the thing to actually work.

The rocket is simple in comparison to changing the systems of an organisation. How much restructuring and reorganisation can an organisation take and still be the same organisation - the same system? How often do you hear organisations described as "it is not the same place as when you were there"? Taking the view that the organisation is a socio-technological device, like a ship or even an orchestra, then by that description, names of the crew members, or players, are not important. In other cases the social aspect of the organisation is everything. There are many perspectives by which something is seen to change. The system point of view defines things as issuing from the statement of purpose, the requirements. Certain subsystems can be changed as long as the overall capabilities remain the same. The statement of capabilities or intention is the mission statement. These have become very fashionable as a type of statement of intention and motivation. (They are always noble.) It is usually taken as the most general statement of purpose governing the whole system. It might, but need not, imply statements of purpose for smaller scales of subsystems such as the sales or human resources subsystems. If it can be read to imply something about subsystems, it will usually be in meeting some measurements - some performance indicators: sales budgets are to be met, human resources will maintain high levels of employee satisfaction. This gives a reason for control parameters in the details of the subsystem. It also gives a sense of relevance to various performance measures. The problem is that the more social processes that are involved in the changing of systems, the clearer the description of the changes needs to be to avoid legal wrangles.

This leads to a detailed description of how to proceed from the inception of the project to the final delivery. This has to be agreed with those receiving the device or system and everything is written down and signed. This is the systemisation of the creation and deployment of systems. It is known by a variety of names but generally under the rubric of Quality Management Systems. The first axiom of a quality management system is that there is a process which, if followed, leads to the required system, free of defects, being delivered to the customer at the agreed time and cost. The axiom does not preclude defect free systems being delivered in some other way. A second axiom, similarly unstated, is that of the analytic method. Everything is planned down to the last detail. Everything can be broken down from the largest scale to the smallest relevant scale. The whole is perfectly explained by the sum of its parts. Making the system requires that the plan is worked in reverse. Small details accumulate into larger components which in turn become subassemblies and so on until the system, device or factory, is ready for testing. The third axiom is that everything that can be tested before delivery will be tested. As noted earlier, some things - such as bridges, buildings, motorways, dams and even organisations - cannot be tested. They work or they don't. But every component which can be tested will be tested. This is quality control and is only part of the bigger picture of quality management.

Testing the components themselves, such as valves and ignition systems of the rocket engine, and the assembly of components - the whole engine - requires co-ordination of the design, construction and assembly throughout each level of the product. It is worth looking at this aspect of systems because that is where the small details affect larger components and determine the success of a system or not. A successful system is like an orchestra - all parts work together to fulfil a purpose. Failure only requires the failure of one small piece to ruin everything. A tragic example of this is the case of the frozen O-rings, which caused the space shuttle Challenger to explode in 1986. The solid fuel booster rockets were designed in cylindrical segments which then were slotted together. Unfortunately, these rockets flexed outward when burning. Strong heat resistant rubber O-rings were fitted in the area of the joined segments to flex with the rocket and stop hot gases escaping. When cold, these O-rings lost their ability to mould into the gaps caused by the flexing. The assumption was made on that fateful frosty morning that the heat of the rocket engines would warm them sufficiently rapidly that they would seal the gaps. The assumption turned out to be wrong, hot gas escaped, and burned into the main liquid hydrogen tank causing the shuttle to explode. None of the O-rings actually broke; one of them just failed to perform its intended function. It did not even fail expectations because it was never expected to operate in frosty conditions. After all, frost in Florida is a rare occurrence.

In physical terms the shuttle is only an inventory of parts bolted, welded, glued and otherwise stuck together. The shuttle and its large components are linked in our minds with specific purposes. To make it work we have to attend to the details of physical things.

Systems Intolerant of Errors - a catalogue of disasters

Unlike computer systems, physical, mechanical and chemical systems are associated with safe work practices. Most components have maintenance requirements. We are used to this with cars. The tyres must be maintained close to a specific pressure and depth of tread, the oil must be changed, and so also with the timing belt, air filter and so on. Even in the modern infrastructure of cities, the piped water supply requires constant maintenance as does the sewage system, the power, gas, transport and telephone systems.

The more interrelated the components of the system, both large and small, the more care - the more detailed and stringent - the safety practices.

Such safety practices might seem irksome and overly cautious until something goes wrong. The following cases show how absolute devotion to specified work practices is important for the large systems which give the shape of the modern systematised social infrastructure.

Piper Alpha

The catastrophic end of the Piper Alpha oil drilling platform in the North Sea provides a frightening illustration of the need to attend to all levels of details.

It was built in 1976 for oil processing but later modified to process gas with major gas processing functions sited near the platform's control centres.

A work permit was raised for a faulty pressure valve. The valve was part of a pump for which a two week maintenance order was raised. On July 6th 1988 the pump malfunctioned. The two week maintenance order was found and work started immediately. But the work was not finished by the end of a shift and a temporary plate was put over the open segment of the pump. The work order was not signed off. That area of the oil rig was restarted on the assumption that the work had been completed. When the pump restarted the temporary cover plate on the valve blew off starting a fire.

Of the 226 people on the platform, 167 died. The rest survived by jumping 10 storeys, roughly 100 feet, into the North Sea, an act absolutely against all procedures.

A number of factors might have mitigated the disaster. The automatic firefighting equipment was switched to manual as maintenance divers were operating near its sea water intakes. It should have been switched back to automatic as the divers only worked during the day and the fire started at night. When the fire was started the firefighting equipment switch couldn't be reached (or was never activated). Piper Alpha was part of a network of oil pipes linking the Tartan and Claymore oil platforms which should have shut down production when they saw the fire and heard the mayday call. They were reluctant to do so because once shutdown, restarting is a very lengthy process and interrupts production. Oil platforms only pay back their supporting investors and bankers when they maintain a near peak production. Keeping the production going kept up the pressure of gas which was feeding the fire.

Bhopal

(The following account is extracted from TED, the Trade and Environmental Database available from:
http://www.american.edu/projects/mandala/TED/class/all.htm)

According to many, Bhopal is the site of the greatest industrial disaster in history. Until 1979, the Indian subsidiary of Union Carbide used to import MIC or methyl isocyanate from the parent company. After 1979, it started to manufacture its own MIC. MIC is one of many "intermediates" used in pesticide production and is a dangerous chemical. It is a little lighter than water but twice as heavy as air, meaning that when it escapes into the atmosphere it remains close to the ground. It has the ability to react with many substances: water, acids, metals, and the small deposits of corrosive materials that accumulate in pipes, tanks, and valves.

On the night of December 23, 1984, a dangerous chemical reaction occurred in the Union Carbide factory when a large amount of water got into the MIC storage tank # 610. The leak was first detected by workers about 11:30 p.m. when their eyes began to tear and burn. They informed their supervisor who failed to take action until it was too late. In that time, a large amount, about 40 tons, of Methyl Isocyanate had poured out of the tank for nearly two hours and escaped into the air, spreading within eight kilometers downwind, over the city of nearly 900,000. Thousands of people were killed (estimates ranging as high as 4,000) in their sleep or as they fled in terror, and hundreds of thousands remain injured or affected (estimates range as high as 400,000) to this day. The most seriously affected areas were the densely populated shanty towns immediately surrounding the plant - Jayaprakash Nagar, Kazi Camp, Chola Kenchi, and the Railway Colony. The victims were almost entirely the poorest members of the population.

This poisonous gas caused death and left the survivors with lingering disability and diseases. Not much is known about the future medical damage of MIC, but according to an international medical commission, the victims suffer from serious health problems that are being misdiagnosed or ignored by local doctors.

The immediate cause of the chemical reaction was the seepage of water (500 liters) into the MIC storage tank. The results of this reaction were exacerbated by the failure of containment and safety measures and by a complete absence of community information and emergency procedures.

A listing of the defects of the MIC unit runs as follows:

153

-Gauges measuring temperature and pressure in the various parts of the unit, including the crucial MIC storage tanks, were so notoriously unreliable that workers ignored early signs of trouble (Weir, pp.41-42).

-The refrigeration unit for keeping MIC at low temperatures (and therefore less likely to undergo overheating and expansion should a contaminant enter the tank) had been shut off for some time (Weir, pp.41-42).

-The gas scrubber, designed to neutralize any escaping MIC, had been shut off for maintenance. Even had it been operative, post-disaster inquiries revealed, the maximum pressure it could handle was only one-quarter that which was actually reached in the accident (Weir, pp.41-42).

-The flare tower, designed to burn off MIC escaping from the scrubber, was also turned off, waiting for replacement of a corroded piece of pipe. The tower, however, was inadequately designed for its task, as it was capable of handling only a quarter of the volume of gas released (Weir, pp.41-42).

-The water curtain, designed to neutralize any remaining gas, was too short to reach the top of the flare tower, from where the MIC was billowing (Weir, pp.41-42).

-The lack of effective warning systems; the alarm on the storage tank failed to signal the increase in temperature on the night of the disaster (Cassels, p.19).

-MIC storage tank number 610 was filled beyond recommended capacity; and a storage tank which was supposed to be held in reserve for excess MIC already contained the MIC (Cassels, p.19).

Cassels, Jamie. The Uncertain Promise Of Law: Lessons From Bhopal. University Of Toronto Press Incorporated. 1993.

Weir, David. The Bhopal Syndrome: Pesticides, Environment, AndHealth. Sierra Club Books, San Francisco. 1987.

Other notorious industrial accidents include:

Flixborough, England, 1974: cylohexane explosion; 28 killed, 89 injured, 3,000 evacuated, 60-acre site completely destroyed.
•Seveso, Italy, 1976: dioxin escape; many children disfigured by chloracne, 4,450 acres of farmland poisoned, 100,000 grazing animals killed, 1,000 people evacuated.
•Montana, Mexico, 1981: chlorine release; 29 killed, 100 injured, 5,000 evacuated.
•Basel, Switzerland 1986: 66,000 pounds of pesticide leaked into the Rhine.

Nuclear Power Plants

Browns Ferry 22 March 1975. A technician checking for air leaks with a lighted candle caused $100 million in damage when insulation caught fire at the Browns Ferry reactor in Decatur, Alabama. The fire burned out electrical controls, lowering the cooling water to dangerous levels, before the plant could be shut down.
Three Mile Island 28 March 1979. A major accident at the Three Mile Island nuclear plant near Middletown, Pennsylvania occurred at 4:00 a.m. after a series of human and mechanical failures nearly triggered a nuclear disaster. By 8:00 a.m., after cooling water was lost and temperatures soared above 5,000 degrees Fahrenheit, the top portion of the reactor's 150-ton core collapsed and melted. Contaminated coolant water escaped into a nearby building, releasing radioactive gasses, leading to as many as 200,000 people being forced to flee the region. Despite claims by the nuclear industry that "no one died at Three Mile

Island," a study by Dr. Ernest J. Sternglass, professor of radiation physics at the University of Pittsburgh, estimated that the accident led to a minimum of 430 infant deaths.

(These accounts are from TED = Trade and Environmental Databasehttp://www.american.edu/projects/mandala/TED/class/all.htm)

Chernobyl

The Chernobyl nuclear reactor disaster has a different lesson. Large systems have many modes of behaviour which excite the curiosity of knowledgeable operators. They are not the things which can be tested. To experiment with them can lead to a disaster. (This account is an extract from: http://www.american.edu/projects/mandala/TED/CHERNOB.HTM)

The genesis of the now infamous explosion at Reactor Four of Chernobyl nuclear power plant began only a day before the actual explosion, on April 25, 1986. On April 25, plant technicians began to reduce power levels in the plant in order to run an experiment with the reactor's main turbine. The plant's technicians wanted to see if, in the event of a power cut, the declining momentum of the turbines could generate enough electricity to power the pumps for forty or fifty seconds before the standby diesel generators took over. Early on April 26, the actual experiment began by stopping power to the main turbine. The flow of the water that normally cooled the reactor was reduced and certain safety devices were disengaged. The reactor immediately began to overheat dangerously, but since the emergency cooling system had been shut off some twelve hours earlier, there was no backup. Within seconds, there was a tremendous power surge that caused two explosions, blew the roof off the reactor building and ignited more than 30 fires around the plant. The damaged reactor core and the graphite surrounding it began burning at temperatures as high as 2,800°F.

The tremendous explosion from the reactor sent a radioactive "plume" of radionuclides into the upper atmosphere, where they began to drift westward towards the rest of Europe. According to Segerstahl "from 26 to 28 April, a high-pressure area over north-east Europe carried the plume northward, at first affecting the U.S.S.R., then later north-east Poland and Scandinavia where radiation monitors in Sweden and Denmark indicated abnormally high readings. The triggering of these monitors was the first indication in Western Europe that a significant nuclear accident had occurred.

Soviet firefighters using military helicopters finally managed to extinguish the blaze in reactor four by dumping between 5,000 and 6,000 tons of boron, lead and other materials onto the reactor core. Twelve days after the accident, the final fire was extinguished. In addition to fighting the blaze, the Russians began evacuating villages within close proximity to Chernobyl within 36 hours after the blast, including the entire town of Pripyat, the city closest to the Chernobyl plant. According to Zhores Medvedev, the total number of towns and villages evacuated was 186 (2 towns and 184 villages), some as far away as 80km west, north and north-west of the reactor site.

The total number of people evacuated from these towns runs as high as 180,000. Once the reactor fires had been extinguished, the Soviets began the work on entombing reactor four. They did this by constructing a cement and lead "sarcophagus" around the reactor. The Soviets accomplished this task using a combination of manually operated cranes and robot machinery to construct the massive sarcophagus structure around the reactor. By November 15, 1986, the Soviet newspaper Pravda reported that construction of the sarcophagus had been successfully completed.

Aircraft disasters are potent teachers of what it takes to create our extraordinarily reliable, but not infallible air transport system. It is a mantra of continuing improvement, dogged investigation of every crash and speedy consideration and implementation of all suggestions to improve safety.

The following descriptions are taken verbatim from the reports associated with the listing of aircraft accidents on http://dnausers.d-n-a.net/dnetGOjg/Disasters.htm.

McDonnell-Douglas DC-10-10 crash at Sioux Gateway Airport, Iowa

(The following is the executive summary of Report No. NTSB-AAR-90-06 Report Date: November 1, 1990 Pages: 126)

On July 19, 1989, at 15:16, a DC-10-10, N1819U, operated by United Airlines as flight 232, experienced a catastrophic failure of the #2 tail-mounted engine during cruise flight. The separation, fragmentation, and forceful discharge of stage 1 fan rotor assembly parts from the #2 engine led to the loss of the three hydraulic systems that powered the airplane's flight controls. The flight crew experienced severe difficulties controlling the airplane, which subsequently crashed during an attempted landing at Sioux Gateway Airport, Iowa. There were 285 passengers and 11 crew members onboard. One flight attendant and 110 passengers were fatally wounded.

The National Transportation Safety Board determines that the probable cause of this accident was the inadequate consideration given to human factor limitations in the inspection and quality control procedures used by United Airlines' engine overhaul facility, which resulted in the failure to detect a fatigue crack originating from a previously undetected metallurgical defect located in a critical area of the stage 1 fan.

The fan disk was manufactured by General Electric Aircraft Engines. The subsequent catastrophic disintegration of the disk resulted in the liberation of debris in a pattern of distribution and with energy levels that exceeded the level of protection provided by design features of the hydraulic systems that operate the DC-10's flight controls.

The safety issues raised in this report include:

1. General Electric Aircraft Engines' (GEAE) CF6-6 fan rotor assembly design, certification, manufacturing, and inspection.

2. United Airlines' maintenance and inspection of CF6-6 engine fan rotor assemblies.

3. DC-10 hydraulic flight control system design, certification and protection from uncontained engine debris.

4. Cabin safety, including infant restraint systems, and airport rescue and firefighting facilities.

Recommendations concerning these issues were addressed to the Federal Aviation Administration, the Secretary of the Air Force, the Air Transport Association and the Aerospace Industries Association.

This antiseptic summary gives no account of the desperate, but cool and skilful, handling of the aircraft by the crew who were very close to making a safe landing when, seconds before touchdown, a wing tilted into the runway and all was lost. The crew only had engine power to control the aircraft.

The next two disasters show that the point of the disaster need not be a failure of the aircraft, or other system, that is destroyed. Aircraft are nestled in an array of supporting systems which contribute to their safe operations.

The Case of unauthorised cargo

On Saturday, May 11, 1996, about 14:15 EDT, a Douglas DC-9-32, N904VJ, operated by Valujet as flight 592, crashed during an uncontrolled descent from 10,000 feet. The 2 flight crewmembers, 2 cabin attendants, and 106 passengers were fatally injured. The weather at the time of the accident was VFR (clear of cloud). An instrument flight rules (IFR) flight plan had been filed. The flight was being operated under 14CFR Part 121. The flight departed at 13:44 after being delayed because of weather in the Atlanta area. Shortly after takeoff, the first officer radioed Miami Approach and requested an immediate return to the airport because of "smoke in the cockpit, smoke in the cabin." During the return to the airport the aircraft crashed in an isolated portion of the Florida Everglades, about 18 miles north-west of the airport. Preliminary examination of the wreckage and interviews with employees of Valujet and a contract maintenance company (Sabretech) indicate that a fire erupted in the forward cargo compartment. Additionally, it was determined that oxygen generators, used in aircraft passenger service units and classified as "HAZMAT" were on board flight 592 and had been loaded in the forward cargo compartment.

The fact that the oxygen generators had been sent back to the main Valujet depot was a result of a series of mistakes and, finally, Valujet was not an authorised carrier of hazardous material. The cargo should never have been loaded.

Crash on Mount Erebus

In the 1970s Air New Zealand started tourist flights to Antarctica. These flights, in DC 10s, did a round trip between the Ross Ice Shelf and Christchurch in New Zealand. Air New Zealand as a tiny operator on the world stage was proud of its well earned reputation as an accident free world-class carrier. In one day it was demolished. On 28 November 1979 ZK-NZP crashed into Mt. Erebus in Antarctica, killing all 257 on board. The only adverse conditions were mists below the aircraft and the cloud base of about 3,500 feet. This led to a decision by the pilot to descend below the cloud base and so the aircraft was flying level at 1,500 feet when it crashed onto Mt Erebus. White mist below, a white landscape and white cloud above produced the strange phenomenon of "white-out", a diffuse shadowless illumination and a mono-coloured white surface which washes out all the usual features by which one can orient oneself. In particular, rising land is not obvious. When the slopes of Mount Erebus became detectable and the automatic pilot sounded "Pull Up Pull up", a five second delay proved fatal. Tragically, 15 seconds before the signal to pull up, the pilot announced they would "have to climb out of this". The pilot was confused because the navigation instructions had only been corrected the night before on the computerised system which issued the navigation co-ordinates. For the previous fourteen months, a data entry error had given the wrong longitude for Mt Erebus, an error of about two degrees. Although the error had been discovered two Air New Zealand tourist flights earlier, it was not corrected until the night before the ZK-NZP flight. The error had not mattered for the military and scientific flights to Antarctica which flew routes designed to give Mt. Erebus a wide birth, and all previous tourist flights had not had white-out problems and had been able to identify features in the landscape.

Systems to Control Systems and so on to Ignorance

Are there any principles which we can draw from these tragedies?

In terms of systems, the answer is in favour of the systems. Certainly they occasionally are the source of their own disasters, but overall it is the social milieu in which they are imbedded that is frequently short of the requirements of the system. According to Norman Simenson, one of the pioneers of real-time software (designed to control physical processes) and editor of the Federal Aviation Administration's Software Interface Group's *Interface* newsletter, "Engineering disasters are always the result of bad management and never the result of bad engineering - or almost always.".

(http://www.stsc.hill.af.mil/crosstalk/1998/apr/disasters.asp.)

In spite of the publicity given to air crashes, its safety statistics continue to improve impressively. Its engineers, designers and safety officers do learn from the past. Each new generation of airliner is safer than the previous generation. From about 1970 to 2000, the accident rate reduced to one fatal accident per 140 million miles flown to one per 1.4 billion miles. Further reinforcing the management rather than system culpability of accidents is the statistic that 80 percent of the world's accidents are caused by 20 percent of the airlines (independent of distances flown).

As much as we might suspect systems, this is a powerful vote in the overall professionalism of the creators of systems. If this is the case, there is a powerful impulse to search for improvements which invoke systems. We shall discuss two related approaches. The first is to look very closely at the system as an analytic object and include even more things which alert human operators that things are going wrong. The second is to expand the area in which the system manages itself. Both of these approaches will be taken in various degrees. In both cases there will be diminishing returns as a great deal of money is expended on controlling events which are extremely unlikely ever to occur.

Even with the best will in the world, things can go wrong because an invisible flaw can cause a disastrous chain of events. As mentioned before, a physical system is just that - an arrangement of matter designed to fulfil a purpose. We design to that purpose and by carefully matching the way subsystems contribute to the overall purpose, layer upon layer of subsystems are accumulated down to the physical bedrock - the actual device.

To link any two adjacent levels of the system, some precise relationship must be given between the inputs and outputs of the two levels. There are always changes in the operational levels of the inputs and outputs which can go undetected at the higher levels. There is nothing we make which can record all the possible changes in the inputs of the system and the materials which constitute the device. Tiny fractures occur in metals, contamination and impurities occur in all inputs, temperatures cannot be guaranteed to be even throughout a material, concentrations can vary. These variations which escape notice or measurement are information which is lost. Variations which are assumed not to alter the performance of the system constitute the unmeasured *entropy* of the system. It is always a hope that the car, airliner or train in which you travel, has no underlying fault which can undermine the stability of the device. Engineers try to quantify the chances of flaws in material causing disaster. Usually this is not possible and one is left with little more than an estimate that such threatening states are isolated, infrequent and have little chance of happening. These are the "one in a million" chances. The more complex the device or process, the more likely the unmeasured variations will be numerous. This is particularly so if there is a big difference in the number of inputs and outputs. This is why baking a cake can fail more often than cooking toast, and why starting a new business in a competitive environment is more difficult than building a house. The number of things which can go wrong is much larger in the first case. It is also true of control systems for nuclear power stations, chemical factories, oil refineries, and early warning systems for semi-automated missile firing systems.

The lesson of the shuttle is that a perfectly good product can be delivered which will work well in specified conditions. But that is only the start. As mentioned above, the whole thing can never really be fully tested. This is because the purpose of a very complex one-off project transcends the sum of the testable parts. It is also because it comes with operating instructions or work practice disciplines which properly should be regarded as part of the system. These are the ways people should operate the system if they want it to run safely. We can ask how do we know that the delivered item is appropriate: that not only does it work but also that it is workable in the explicitly stated conditions, where these are known. And something the Challenger space shuttle established is that we don't always act on what we know we don't know.

The answer is that for large special purpose devices, one cannot tell how workable and in what range of conditions. Particularly one cannot tell whether the operating instructions impose expectations of good behaviour which are too extreme. All that the axiom of testing can claim is that testing has occurred on all levels. Typically testing starts with unit testing. This is testing a component on its own. To illustrate this, think of an commuter train. The electric doors can be considered "a unit". Unit testing tests opening and closing the door on its own before it is fixed into the carriage. The signals to open and close are simulated by general purpose signalling equipment. The unit test will test such things as whether the doors stick or squeak and that they open fully and will not close if they detect something between them. If the unit test is satisfactory, the doors are now fixed into the carriage as they will have to operate. A set of tests is again run to make sure the device works in its appropriate context. This is called integration or system testing. It is a test of the unit in its full environment. Then there is the user acceptance test. The train goes on a journey and people get on and off through the door. This test tests that it stays open as it should, closes appropriately, and fulfils its general expectations. System tests are sometimes called factory tests and acceptance tests are called site tests. Integration testing is really the start of the testing of the system. Individual components might all work extremely well. They can come through all sorts of unit tests with flying colours. Hardy little beasts which never fail. Putting them together, however, might produce an antisocial mob which is difficult to control. They fail the integration tests in all sorts of nasty ways. Rather like individuals turning into a mob, they amplify the nasty corners of each other's behaviour. It is very difficult to predict how the interaction of not just one, but many interacting devices, will emerge. And yet we are astonishingly good at this type of physical engineering. This sequence of testing is applied to everything from new toasters to nuclear powered aircraft carriers.

There are various other tests. For example, when an upgrade of some component is fitted, such as a carburettor in a car, the obvious thing to do is to go for a drive and check how well the car runs. In a formal situation this is called a regression test. But one thing which is not tested is how well it works, in interaction with other systems, with an averagely trained operator. Cars might be taken to fail this test.

The cases of the space shuttle Challenger and the DC10 which lost a chunk of its turbine, show that the components, or units, can fail, initiating a tragic set of events. How can we avoid this? It is doubtful that much could have been done with the Challenger given it was outside the known operating limits of one component. What of the DC10? What amount of pre-flight checking or additional on-board checking devices are worth it.

Every new device devoted to merely monitoring equipment adds weight and interacts with the rest of the equipment. The very first space shuttle had four computers which polled each other and took votes on discrepancies. This was to ensure the shuttle did not depend on just one computer. It was "multiply redundant" or "overdesigned". Someone felt four was not enough and a fifth computer was added. It was this that caused the delay on the launching pad. It was disagreeing with the four others. The problem of trying to second guess everything that can go wrong is that it changes the nature of the device. It is also very difficult to draw a line and say "well it is this safe at this stage". For the DC10 there is *now* an answer. Following the Sioux City disaster the control of aircraft through engine power alone was investigated. Within six years of the disaster, propulsion controlled aircraft (PCA) control was successfully tested. PCA systems packaged as an addition to fly by wire systems could be fitted to most modern aircraft. But total hydraulic failure hasn't occurred since the crash at Sioux City and the motivation to add this system has yet to overcome its cost.

Reliability engineering takes a statistical approach to shrug off our ignorance about systems. It takes the approach that the customer will accept a certain rate of annoyances - say one a month. At the beginning of a testing program it collects statistics on the rate at which defects are being found and corrected. On the basis of a statistical model it then predicts that the acceptable level of defects will be reached by a certain time. At that stage it is expected (or hoped) that the product will be ready for delivery. This is a realistic approach with many devices and systems. No one can guarantee that new technology is defect free. The time taken to establish the boundaries of operating conditions alone cannot be sharply defined. But reliability engineering can aid in a realistic estimate of good operating conditions and practices.

Another approach to the pursuit of safety is to take as much of the management away from the dangerous parts of the system and put it into a larger "containing" system. This is the idea of a hierarchy of systems, mentioned earlier in the section on science of systems. Thus the decision whether to launch the shuttle would not have been taken by public relations conscious managers, but a launch management system which would have said, "No - too cold". Such an automated management decision system would have a very full model of the dangerous "core" system. It would monitor it closely and manage as much as possible the maintenance of it. It would shut it down when it detected problems in the surrounding environment. Up to a point this is possible. The network of oil rigs connected to Piper Alpha could well have been controlled this way. But again where do we draw the boundary? The management decision system itself would have to monitor parts of itself to know that it is working well. That system would have to be represented in the second decision system. This would be a self-referencing system. In the limit of this "Russian Dolls" of systems, the system becomes fully self-representing. A famous theorem in mathematics, Godel's Incompleteness Theorem, would say that such a system can have states about which it cannot decide what to do. But even at the level of a limited number of dolls in the "Russian Doll", the system will be extremely difficult to understand and, perhaps, only be understood by a priesthood of long serving devoted specialists. Even they would not know all possible actions of the system just as a champion chess player only knows a tiny fraction of all possible games.

The very size of systems will force them to contain tools to aid their own management. Inevitably some degree of self-referencing will be necessary. The larger and more complex a system becomes the greater the necessity to design it in terms of "modular" subsystems: easily separated and identifiable subsystems which are individually replaceable. Whether human or machine these are represented as agents. Faulty modules, or agents, can be replaced, which usually leads to a nice piece of international trade. Just as the use of an infrequently used feature of a word processing package can hold one up, so also unfamiliar problems in a production or control system can lose a great deal of production or even threaten the workings of the system. The larger the system the more frequently this can be expected to happen because ignorance of a system grows with its size. This is circumvented by adding a systems encyclopaedia to the system. This might include an expert system, a system which simulates the live system and the usual "hypertext", or linked and cross-referenced, documentation. The more a system manages itself, the greater the requirement to become "self-documenting ". This makes no sense if the system does not change. Once documented no further work is required. But complexity forces a different approach. Each change needs to be automatically registered and documented accordingly. Changes require a statement of purpose which is picked up by the self-documenting algorithm.

The inclusion of self-monitoring and self description is the start to a new extension of systemisation. It is an extension of the idea of self-correcting goal-seeking systems which have been around since the spinning governor controlled the release of steam in steam engines in the nineteenth century. Systems as diverse as inertial navigation, radar tracking, guided missiles and automated quality systems in manufacturing, use self-correction and goal-seeking. Self-description, and self-reference in general, occur when the system monitors its own logical structure. In such circumstances it can register, report on, and possibly block changes. Self-referencing can apply to goal-seeking itself. Certain operational levels can accept changes of structure and logic as long as the outputs change in some desired way. This builds purpose into the system. A fully fledged self-referencing system could deploy batteries of genetic algorithms, represented by sets of functions with additional variables. When existing functions are failing to deal with trends - as evidenced by increasing errors - functions could be drawn from the store of functions and fed into the genetic algorithm processor to "seed" an improved algorithm for solving a problem. The system monitors the changes and can answer questions about its new self. So much expertise is held within the system that only well trained experts can give a level of service above that which the system itself delivers. Given via a telecommunications link, this makes self-referencing systems the source of very profitable servicing and consulting.

Self-referencing systems are not available at the time of writing. We are still struggling with the level of complexity of control systems of large devices. The integrated control of highly automated factories, power plants and aircraft represents humanity's most extensive application of logic. There are layers of systems encoding the logic and mathematics of machinery and processes which must be integrated. There is no well worked out theory which can deal with the self-description of large-scale control systems. A complication of any system which mixes control and operations is the requirement to deal with tendencies. Describing changes to control based on time series, something common in engineering and becoming common in investments, forces the logic of description to deal with different scales of phenomena and complicates the logic of the descriptive system. As already mentioned, the standard problem of large scale systems is the difficulty of verifying that testing has been completed. This problem becomes acute with a self-describing system. In some areas it could turn out to be an awful liar: an unnerving prospect in aircraft or chemical refineries!

Whatever approach to safety and reliability is taken, systems and regulatory safety schemes in which they operate, ultimately rely on professionals. The specialists who are charged with the good operation of the system. These are the people whose reputations live and die with the system. These people are not optional. They cannot be contracted at the last moment when things are going wrong. They are the boundary between the device and its social use. Organisations frequently regard the maintenance of a computer system, a factory, a fleet of ships, as not being their "core" business. Maintenance is not "core"; service or utilities delivery or products are core business. Give the non-core business to some contracting firm and put more money into marketing. This late twentieth century fashion can ruin an organisation. NASA, the apogee of high-tech excellence, has probably lost much more through the failures of space probes and rocket launcher explosions than it has gained by adopting a commercial "faster, better, cheaper" philosophy. E. M. Hanna, an aerospace safety consultant, found that cost cutting, short-term objectives and the replacement of older, more experienced workers by younger people have resulted in a reduction of critical skills. Adding to this is the new breed of commercially oriented managers who do not want to discuss problems - who have a 'shoot the messenger' attitude. A disaster when everything is a grand experiment.

Less spectacular disasters have been linked to work being done by contractors who were chosen on the basis of quotes rather than carefully assessed expertise. A number of Australian disasters: a mishandled explosive demolition which killed a young bystander, a fire in a naval vessel which killed four, and a gas plant explosion which killed two and injured eight others, prompted a report on the commercialisation of the Australian government public sector. This report stated, "...the committee is extremely concerned at the effect of the public sector's loss of skills and the belief that generalist (managers) on their own can effectively manage technical operations. There is clearly a need for managers to keep a closer watch on the expertise available for carrying out core activities as agencies are commercialised." This theme, a replay of NASA's lament on lost expertise, has been familiar for too long and runs counter to the analysis of the good operations of systems.

Managing the Unlikely and Unruly

Fifty years of computing, financial systems - including foreign exchange - EFTPOS with its variations, and travel booking systems, have added many conveniences to day to day living. The Internet and e-mail are international systems now in routine use.

These are huge systems so what is the problem? No problem - it is just that what we have is fifty years of evolution. How many projects have been buried in history in the last fifty years? How much money is wasted? The Internet is an enormously large system which has simply "evolved" without (much) management and so can be claimed to be an enormously successful large project. But the Internet evolved without a budget and without specifications from the original budgeted and specified Arpanet. The Internet is essentially a horizontal system, and unlike "vertical" systems, does not have much logical depth beyond the "browsers". The programming "team" of the Internet is huge, its time-scale well outside that of specified commercial systems.

Large systems, including personal computer operating systems, and systems which control devices on a grand scale such as the "Taurus" System, the Denver, Hong Kong and Kuala Lumpur airport baggage handling systems, have very poor statistics for being completed on time or on budget. By some accounts 40 percent of large projects fail in some way or are cancelled. But the expectations for very large systems continue. In the 1980s the "Starwars " system was one case in which the public had a clear idea of what testing might look like. It was doomed from conception even though it was cynically supported to keep the research funds flowing. Success of the large projects comes down to good management. There are text books written on project management and it seems dull stuff for the technical people who create the systems. While we need very much improved methods for developing large systems, there is no substitute for project "generals" - gifted individuals who can hold the project together, instil responsibility and urgency, and keep disparate parties communicating. Such project generals are rare. Project management, in terms of meticulous, or rigid, form filling from a crew of gifted individuals, does not succeed without the psychological boost of being a well motivated, well led, "champion" team. The history of systems - the very process of systemisation, once a trickle and now a tide - requires leaders of

Andrew Macfarlane

exceptional gifts. The larger the system the more exceptional these people need to be. Which is altogether very unsystematic.

Chapter 6. Irony: The Illusion of Control

How can we use our ingenuity and experience with systems to transcend our fixation on quantity and speed. What new systems are required in this new century to address the problems that still blight the existence of most of humanity? What new systems are required to roll-back the problems of excess?

Knowing what we don't Know

We don't know how we know. The human being is a poor model for information. We have very little understanding of how we understand anything. The information we use to make decisions is a tiny fraction of the information about us. The body itself provides almost no information about its own state. You do not know the state of almost any organ in the body unless it is in bad shape. In terms of reporting by exception, our bodies only report feeling unwell after the invasion of hundreds of millions of micro-organisms or injuries affecting hundreds of millions of cells. Our hypothalamus, pituitary gland, T-helper cells, lungs, liver, ganglia and aorta give no report of good functioning and frequently only implied reports of bad functioning. Our brains are devoted to the perception and processing of the external world. This is in complete contrast to the usual application of information systems within an organisation. These days any executive has access to a vast amount of internal information. In contrast, information about the world outside the corporation relies on the old sources of newspapers, news broadcasts, government statistics and trade journals, with some additional sources such as market surveys and the Internet.

An organisation's immediate external world is its customers and, certainly, obsessive collecting of customer details and reactions has become *de rigueur*. One thing which has enabled humans to flourish is knowledge of their environment - the usefulness of plants and the habits of animals. Companies attentive to their customers, and willing to understand and fulfil their requirements, flourish well. But no company lives on its own efforts. Each depends to some measure on an infrastructure of the economy. The more types of goods traded in the economy, the more companies spin off one another's successes and spawn new products. The guiding principle here is Margelef's idea that an ecosystem evolves until all energy flows are part of some creature's niche. Similarly modern economies are replete with niche dwellers. What information system distributes these large-scale data? It would be invaluable for companies starting new ventures. They could tell the chances of being behind someone in the market, or how long it might be before a competitor enters and attacks the niche. Fragments of this information already exist in newspapers, government publications and the Internet. The problem is to know what is relevant. Without knowing the connections between events, how can we tell if something is relevant?

We have a paradox. Life is full of chance events, chance insights of genius, overlooked clues, freaks of nature and acts of God. These destroy predictions and undo good works. The story of the collapse of Barings Bank is a story of an increasingly desperate young foreign exchange trader gambling more and more of the bank's money which properly followed procedures would have kept from him. His final undoing was a relatively small decline in the Japanese Yen as a result of the Kobe earthquake. The earthquake disrupted the landing of goods at the Kobe port, adversely affecting the manufacturing economy finely tuned to just-in-time supply. The manufacturing segment of any economy is a network of links forming a loosely knit system. The problem is that the effect of abnormal events might not be appreciated for sometime. Our understanding of large systems is not based on clear insights into the overall shape and form of the system but in little bits of which we have some experience or a reasonable model. The problem is not new. In

"August 1914" the historian Barbara Tuchman relates how the Germans had developed a vast plan for rapid mobilisation using a complex network of railways and train timetables. This was a mobilisation "megamachine ". The Kaiser asked the officer-in-charge, General von Moltke, to set it in motion. The following day he asked for it to be put into action only on the Eastern front, figuring, wisely, that Germany could not win on two fronts. Von Moltke said it could not be done - it was too late, the machine could not be turned off. There has been much argument as to whether Von Moltke was correct. But who at the time could argue? There were no experts in the behaviour of such a mobilisation machine. And it was only turned on once.

In the end humans can only cope with so much information. Prejudice favours certain parts and discounts others. When this process is successful we put it down to the ability to cut through the irrelevancies and go to the heart of the problem. Call it intuition, experience or ineffable insight, it is indicative of *not* using information. The best tested technique for handling a mass of information about interconnected events remains the ability to abstract out the interactions which dominate. If these are few and simple we have the basis for a model of a system. If not we are frequently forced into complex models which simulate the phenomena. Unfortunately the resulting simulation may not reveal much that we can understand. A simulation of thousands of coloured point 'objects', which interacted and changed colour according to simple rules, produced no clear pattern for thousands of cycles. Then spirals of colour started to form and rotate. What types of abstractions will deliver an explanation? What types of social phenomena - with their complex rules of interaction - appear after hundreds of complex interactions? Even if theories exist, they are notoriously hard to check let alone think of practical experiments that might prove them false. Examples of this are the various experiments in "computer life". These experiments are simulations of various rules of interaction which are supposed to mimic biological "rules". They can be seen as pioneering efforts in computable biology. What they represent, and whether they resemble anything in the real world, depend on unprecedented levels of inspired interpretation.

On more modest scales, endless spreadsheet models purport to capture some aspect of the world. They allow numerous "what-if" calculations. Unfortunately, as complexity increases, the results require more and more expertise to sift and judge. Time alone does not test the models as the models assume a set of conditions which usually does not endure. This is true of large reporting systems as well. Database systems which are the work of scores of people over years are usually so unwieldy that no one person knows all the details. Specialists might gain confidence in certain parts but information collated from scattered parts of the system requires a modicum of faith to accept it. This is faith in a system which is only understood in terms of generalities. Of course we do this everytime we transfer money internationally, use our credit cards, fly in an airliner, or use a train timetable.

However we model the business or organisational world, we remain the final arbiters of whether the results of models, projections and reports are worth having. We rely on experience and expertise to make decisions. We select data and conclusions which "make sense" to us and query data that doesn't. We have to do this; as humans we compulsively order the world to give it meaning, and, like all animals, we evolved with our meanings. For meaning is what determines behaviour. We do not rely on systems because we can never hold enough data in our heads and make decisions without emphasising our own, limited, experience. Any guess as to the social consequences of business or political acts is filtered through our culturally derived "commonsense" - the prejudices of our circle of acquaintances.

The computer industry lives on the slogan that more information is better. However, the human is not a vast information processor wonderously drawing conclusions from a mass of data. Evolution has yet to find a use for such a creature. Certainly the most powerful minds can hold a formidable amount of data but they are inevitably shaped by ideas, concepts and models of our natural environment which filter and re-interpret. This reduces to human scale, the use of systems of vast collections of data. Is the Internet a counterexample to this piece of "commonsense"? In 2000 the Internet had 200 million "pages" and resembled a telephone directory, a huge magazine shop and gossip column all rolled into one. Its advantage over a good library is rapid publication and simple searching. But it is a hodge-podge of opinions and advertisements, not a radical new approach to organising knowledge that will break down prejudice and usher in enlightenment. Indeed its relevance to education is that it should be used as a dubious source to develop techniques of checking, sifting and comparing with other data. It has little to do with the development of mental capabilities of anyone especially. It is irrelevant to the development of the capabilities of good judgement, penetrating logic and mental agility which are the valuable mental skills in an age of systems. Of course all this is subject to rapid change. Large corporations such as Time-Warner are forming consortia with groups of universities. World-wide, higher education is expected to be worth $US4½ trillion by 2020 (*The Australian* 22-23/4/2000 quoting an IBM report). This will be a feeding frenzy as education changes to an Internet-mediated skill package. Left to the stuff of nostalgia will be the pleasures of experiencing a passionate teacher and scholar in a social milieu of argument which tests the ability to juggle ideas.

These considerations might temper our personal reactions and attitudes to systems. They do not reduce the momentum of the production of systems. One is assailed by the achievements of systems, particularly information systems. The triumphs of chess programs, of diagnostic expert systems, computer animation in films and our ability to compress prodigious quantities of data on tiny discs give an impression that systems are not only unstoppable but omnipotent. There are, however,

177

factors which derive from our own ignorance which limit the momentum of the deployment of systems.

"Knowledge dwells
In heads replete with thoughts of other men;
Wisdom in minds attentive to their own."
William Cowper, The Task, 1765.

To go beyond the level of systemisation which we enjoy today, and which has been sketched in previous chapters, requires building a level of "intelligence" into a system. Intelligence must include the ability to work with humans in a way that accepts our nature and our type of knowledge of the world. There are a number of artificial intelligence projects which try to build "commonsense " into machines. If this be successful, domestic robots will be possible - obviously an enormous market. In order to program this, we have to model it. In order to model commonsense we have to understand it. In order to do this we have to solve a number of perplexing problems in thinking and the way "commonsense" provides a foundation to knowledge. The problems of understanding commonsense take us straight into the subject of epistemology - that part of philosophy which deals with justifications of knowledge.

The very first problem is processing inputs into recognisable objects such as "cat" or "chair" or "teapot." It might be possible to train neural networks to recognise a vast array of common objects. However, we would want to deal with classes of things such as "furniture" or "vegetables" or "pets." Even if we solve the problem of recognition that a cat shape is indeed a cat, we have the problem of figuring out the "essences " of sets of objects. This problem of essences is part of a subject called phenomenology, a subject associated with the name of Edmund Husserl (1859 - 1938). Should our attempt at using neural networks fail, and we have to work out the essences of the phenomena of a cat, we have a lot of work to do. Describe for yourself the shape of a cat and distinguish it from small fox terriers and dogs with short noses and pointed ears.

The next problem is to get our robot to deal with cause and effect. This is sometimes hard for children but only a damaged child does not know when a ball has knocked a vase off the table and it is time to clear out. The problem is to distinguish close association and cause. After all, the cat might have been on the table, close to the vase when the child threw the ball at it. Distinguishing cause from close association is something that has concerned philosophers ever since David Hume first discussed the problem in the 18th century. A related problem - also associated with Hume, and part of the cause and effect problem, is generalisation - or what is called induction. Commonsense requires that we generalise. Statistical programs deal very well with calculating a curve through a scattered set of points which, by some measure, is the best possible, but this is not what we mean by induction. The association of dog and cat followed by chasing has a statistically strong coefficient of correlation but this is built up over a lot of trials. It's a good start but not much good for a robot servant. It is especially useless when the statistical association is weak.

An attempt to capture the lack of logical definition in statements like "this coffee is too strong for me", is associated with the name *fuzzy logic*. The concept of "strong", as with coffee, has no strict boundaries. In fact fuzzy logic is just a technique for manipulating functions associated with qualities, such as the strength of the coffee, which go from zero when something is not true, to one when it is true. Trendy though fuzzy logic is, it is probably not much use in the understanding of cause and effect. It makes little sense to assign a "perhaps" value of 0.6 to "the cat caused the vase to fall". Such a number can only have meaning in statistical terms anyway.

Commonsense not only involves classifying phenomena and distinguishing cause and effect from close or statistical association, it requires a sophisticated sense of concepts of space and time, quantities, qualities, and concepts such as necessity and possibility. These are all concepts and ideas which the philosopher Immanuel Kant worked with in the late 18th century. Commonsense turns out to be an enormously slippery or sophisticated idea. Kant's analysis of why we have a clear idea of cause beyond close association is both technical and contentious. What is clear is that the human mode of understanding common things - commonsense - is very difficult to describe. And what is difficult to describe can be beyond the limits of logic.

These epistemological considerations are not insurmountable because there is nothing decisive about them. They can point to the difficulties faced by any system which simulates commonsense, or more precisely, human sense. They are under challenge from new techniques in pattern recognition. There are now systems which have succeeded in classifying patterns of information remarkably well. These usually involve artificial neural networks.

The long-term importance of artificial neural networks and genetic algorithms will be to accelerate systemisation. They constitute an important contribution to a new way of finding rules to mathematically express a way of getting a decision. They are not unprecedented in the bag of techniques used by mathematicians. Techniques for casting about among possible options to come up with the optimal ones have existed for generations. But these optimisation techniques have started with a clear idea of both technique and solution. They can independently check a solution with other techniques. The difference with the new techniques here is that an experimental situation has to be set up and the answer proceeds from the experiment. But usually there is no easy way to recheck the results. The generic algorithm might have proceeded down the wrong track and come up with the correct answers in the cases where we have answers. The artificial neural network might have learned just enough to be pronounced a success, but in fact produces crazy results outside the trained responses. While none of this is strictly new, it is a type of knowledge that leaves us uneasy because it is answers without insights. As science it does not advance our knowledge of the world in any clear way.

If the apparent philosophical limits on technique look as though they have been evaded, at least in part, are there limits to the domain of systems? That depends on what you see as a system.

It is no longer possible to separate large systems from the social infrastructure, and we like it that way. We actually like a predictable orderly world. We don't like not being able to trust those with whom we come into contact. We want our modes of behaviour and commonsense to be shared by those with whom we work, play and shop. We like ATMs, telephones, electric power, well sealed roads, drinkable tap water, predictable and plentiful public transport and entertainment. Even those who are vociferous against "The System": entertainers, playwrights and pop-singers, want these things. Furthermore, the well-satisfied middle class wants to move around the world with these systems ubiquitously and quietly serving their comforts. And each generation wants more. This means more connected systems with ever more options in them. Options drive more products to suit "lifestyes". Banking becomes finance management with products tailored to each stage in life. Education will become "product driven" in the same "market led" way. This means systems to administer and deliver the products. The depth and breadth of systems continue.

But how well do we understand what we are doing? Economics predicts no stable point of economic activity. Areas of the world will flourish while others decline. Skills are already subject to fashions in technologies, and are beginning to flow around the world like currency. The systems of comfort are accessible by those who have work but work and trade are the subjects of fluctuations. They have their own tides and weather. These large-scale phenomena defeat systems just as human truculence does on the small scale. The efforts of investors and foreign exchange dealers to discern patterns and deploy sophisticated techniques, including artificial neural networks, show that even tiny fluctuations in systems can be used to amass fortunes or lead to disasters. The world-wide money, commodity and stock markets are hugely systematised; huge amounts of data are captured and distributed internationally within fractions of a second, and rules of proper behaviour are enforced. These are among the largest systems we have yet built and yet they control us. We do not control them. We cannot enumerate their states or control them from any point or even from a hundred points. We cannot test them: there is no site test - no way to regression test. We have to believe in them and go with them. Systems were invented to control variation but our largest systems can only report on it. They have not been designed to control what they monitor. The creators of the system do not know enough about the financial system to control it. No one does. Only conformity to the system and its implicit concepts of proper behaviour allow it to endure. There are those who make profits by pushing these international systems to the boundaries of good behaviour. Those whose status and income derive from the system, try to maintain the proper behaviour. We see this whenever currency speculators clash with bankers during times of unusual fluctuations in currencies.

The general rule that the more that is put into a system, the more variation and options it has to deal with, the less likely it will do it successfully, and the duration of those successes will be shorter. The system, as part of society, is subject to social pressures for change. The pace of social change is not going to decrease, so socially embedded systems will forever be subject to tinkering. "Tinkering and tweaking" systems always happens and it is the last thing to get properly tested and documented. With physical systems, the infrastructure of society, we seem to have become used to their costs and dangers, (even though we remain reluctant to provide the billions of dollars required to maintain sewers).

This volatility of the technology, this restlessness, poses a threat to large systems. As "tinkering and tweaking" continues on a large system, the original specialists drift away. What is known about the system and its connections with other systems, contracts to areas most recently modified. Maintenance and documentation of an old system, say over ten years old, are not career enhancing moves in any technology, especially one with an ethos of hectic development. Long, close experience of any given system is a career risk. Years of specialisation can be devalued by a new fashion. To obtain a sufficient depth of knowledge of a large system, one needs to approach it as a scholar approaches the work of a revered author: worthy of years of study. Computer systems do not have, of themselves, anything like this degree of attraction. Large business administration systems have training courses - enough for 700 days. About the same length of study required to obtain a Bachelor's degree. No one does all the training - there are only specialists. The attraction of being a business system expert or a banking system scholar is the same as that of being an airline pilot. The money is good to very good. The speed of change, and the attraction of recruits to lucrative systems areas, rapidly produces a deflationary effect, killing off the original motivation for study. Thus the very social nature and effect of large systems precludes their being studied sufficiently. Technical advances will simplify things. Possibly, well-defined generic subsystems or "abstract data-types", or "objects" in technical language, can give systems much needed logical coherence. Information systems have long envied the way large systems such as airliners are assembled from standard components, such as jet engines, manufactured to specification by "object" specialists. But the envy has not produced much progress. Even with generic objects, the temptation is to add features on features so that systems grow until their management fails - like the man falling from a skyscraper "it's OK so far". And so it is two generations into this new world.

Systemisation continues on the back of previous levels of systemisation. The length of time it takes to create a new layer decreases. To have got started at all required a lengthy period of order in pre-systematised society. Until the middle of the twentieth century, few non-military systems had the potential for creating widespread disaster. The internationally connected multi-subsystem commercial systems of today are increasingly a flamboyant experiment. Commercial urgency and the flood of new technological opportunities have created an atmosphere of "just do it - don't worry about it". Thus layer upon layer of systemisation encloses the corporate world to become its operating "reality". The complexity of a corporation, as with any large organisation, is reduced to a computer representation: all the "reality" with which its operators have to deal. The computer system of a bank is the system of the bank. A defence system is often only known by its representation on the computer. This problem of only knowing part of the world, the part represented on the computer, makes independent judgement very difficult when things go wrong. In the 1987 stock market crash, the system was the reality. The speed at which changes occurred meant that the people involved had to react without any consideration of consequences and the real worth of things in the world. A similar case in which the system insulates the operators from reality is in control rooms, such as in nuclear power stations.

Some would argue that we have passed the point where systems are understandable and that they are about to become far beyond our control.

The Network of Super-computers: The Ultimate Embrace?

First let us postulate that the computer scientists succeed in developing intelligent machines that can do all things better than human beings can do them. In that case presumably all work will be done by vast, highly organized systems of machines and no human effort will be necessary. Either of two cases might occur. The machines might be permitted to make all of their own decisions without human oversight, or else human control over the machines might be retained.

If the machines are permitted to make all their own decisions, we can't make any conjectures as to the results, because it is impossible to guess how such machines might behave. We only point out that the fate of the human race would be at the mercy of the machines. It might be argued that the human race would never be foolish enough to hand over all the power to the machines. But we are suggesting neither that the human race would voluntarily turn power over to the machines nor that the machines would willfully seize power. What we do suggest is that the human race might easily permit itself to drift into a position of such dependence on the machines that it would have no practical choice but to accept all of the machines' decisions. As society and the problems that face it become more and more complex and machines become more and more intelligent, people will let machines make more of their decisions for them, simply because machine-made decisions will bring better results than man-made ones. Eventually a stage may be reached at which the decisions necessary to keep the system running will be so complex that human beings will be incapable of making them intelligently. At that stage the machines will be in effective control. People won't be able to just turn the machines off, because they will be so dependent on them that turning them off would amount to suicide.

On the other hand it is possible that human control over the

185

machines may be retained. In that case the average man may have control over certain private machines of his own, such as his car or his personal computer, but control over large systems of machines will be in the hands of a tiny elite - just as it is today, but with two differences. Due to improved techniques the elite will have greater control over the masses; and because human work will no longer be necessary the masses will be superfluous, a useless burden on the system. If the elite is ruthless they may simply decide to exterminate the mass of humanity. If they are humane they may use propaganda or other psychological or biological techniques to reduce the birth rate until the mass of humanity becomes extinct, leaving the world to the elite. Or, if the elite consists of soft-hearted liberals, they may decide to play the role of good shepherds to the rest of the human race. They will see to it that everyone's physical needs are satisfied, that all children are raised under psychologically hygienic conditions, that everyone has a wholesome hobby to keep him busy, and that anyone who may become dissatisfied undergoes "treatment" to cure his "problem." Of course, life will be so purposeless that people will have to be biologically or psychologically engineered either to remove their need for the power process or make them "sublimate" their drive for power into some harmless hobby. These engineered human beings may be happy in such a society, but they will most certainly not be free. They will have been reduced to the status of domestic animals."

Bill Joy reflecting on The Age of Spiritual Machines
http://www.wired.com/wired/archive/8.04/joy.html

There is a school of thought, represented by such books as Ray Kurzweil's *The Age of Spiritual Machines,* which goes as follows. The rapid miniaturisation of computers, along with the seeming unstoppable growth in power, will lead us to a world of networked autonomous super-computers - beyond anything we have today (whatever your today is). These will swap information through networks much larger and more elaborate than the Internet. This multi-connected information swapping aggregation of systems will amount to an organism. Individual units of the system - such as single programs - need not be all that clever in order for some super-intelligence to emerge. Each neuron in the brain doesn't do much - it is the fact that there are a huge number of them and they are connected in such staggering ways which fires the spark of intelligence. Extrapolating from the recent history of computing would see a similar vast number of logic elements available for connection. From there our vast network takes on its own life and we had better prepare for this new world order.

The new part of this argument is that cleverness is not needed. Smart algorithms need not apply. Like humble neurons in vast numbers, it is interconnectedness which counts. The whole network - not the individual elements - assumes characteristics on a different scale to the elements.

As applied to neurons, the argument is true. It is also true of individual cells. Each cell, admittedly much more complex than any piece of software at the time of writing, is only important in its tiny place in the whole of the body.

The concept behind the age of spiritual machines is that a group of entities which individually do not do much beyond carrying out simple functions, can coalesce in some type of organisation which transcends their individual abilities. This is seen in biology in which cells aggregate to become an animal. It is also seen in the way ants produce a complex ants' nest, bees a beehive, humans a city. The assumption is that under fortuitous conditions, simple software items, which on their own are not complex, can spontaneously - that is without human intervention - form a structure that is highly complex. The structure is much more than the sum of its parts, just as is the case of animals. Indeed it is anticipated that it will be conscious, intelligent, and even have emotions.

The practical consequence of this is that our acquiescence to enjoying world-wide computer networking - and hence our voluntary social dependence on it - plugs us into something that we cannot readily describe except in terms of analogies. If the network has become conscious and has emotions, what does a bad mood with the world financial transactions mean? What happens with international travel when the network is feeling particularly jolly?

Starting with the idea that the computer components are simple but numerous means starting with simple algorithms. In technical language this argument might be couched in terms of Turing Machines or recursive functions. For our purposes we can think of these as devices which accept some data, apply a calculating rule, an algorithm, to the data and produce an output. Composing Turing Machines so the output of one becomes the input of another is no different to passing the result of one algorithm to another. This produces a new Turing Machine. But we can create something new by adding what the founder of the subject, Alan Turing, himself called "Oracles ". An oracle is something the algorithm does not work out but which is "handed" to it - the answer is given from outside the algorithm.

We can think of an algorithm as a table which relates inputs and outputs. Or, to move to biological language, stimuli and responses. This way it is more obvious how two algorithms can compose. Also, a set of algorithms is still an algorithm as long as there are not two or more ways in which it responds to a given stimuli. If this is the case, it might be that there is additional structure which resolves the conflict. This requires either an oracle or another algorithm which uses additional data to arbitrate. In this way the algorithms are specialised by additional structure.

The evolutions that refine the algorithms occur within larger structures - ones which resolve conflicts. At the same time they restrict the overlap of the algorithms and result in specialisation. The "intelligence" of these additional structures is in the number of oracles they have to produce.

There are an enormous number of unknowns in the way these sets of algorithms can be envisaged to evolve structures.

The first is an analogy with biology. It is unknown that a cell capable of living on its own can be described by an algorithm. If not then the arguments behind the age of spiritual machines cannot use biological analogies. Even the argument that neurons really only output a "fire or not" signal is not impressive. The output of firing a neuron is a sequence of chemical changes, the general effect of which is not fully known. Nor is the communication between different sets of neurons which arise through hormonal signals. Beyond this we have not the slightest idea of how neurons evolved in the evolution of early multicellular organisms.

Suppose for the sake of argument that algorithms can be as complex in processing variable inputs as cells. We have no idea how the oracles specialise the cells (algorithms) over time. We do not even know much about the oracles themselves.

Given the almost complete lack of theory - thus allowing free range to speculation and analogy - what stops the analogy being taken to give the network analogy of a tree or a mushroom (mycelium, spores and all). These are still communities of specialised cells.

Starting with very complex organisms such as bees or ants, creatures which supply a steady stream of scientific surprises, evolutionary pressures - the source of the oracles - need not produce hives or ants' nests. They can produce bumble bees, or that most ubiquitous of species, beetles, with over a quarter of a million species (Wilson, E.O. *The Diversity of Life*, Harvard University Press, 1992, p198).

To produce something like an ant colony requires a remarkable degree of specialisation but not total specialisation. It requires a suppression of variation within the structure. Like an accumulation of expertise in a culture, the evolution of the structure requires an initial variability which feeds back into the culture. Human experience demonstrates this can be maladaptive as well as adaptive. In particular, the structure has no behaviour which is not created from the behaviour of the "cells" or agents within it. It is synchronised behaviour and a chain of actions which produced more than what an individual can do. This is seen in a structured army which achieves more than a set of individual warriors. The team is more than the sum of its parts because its efforts are linked to a purpose. But this purpose is provided by some encompassing system - in terms of sets of algorithms, it comes in the form of many oracles. This is what makes an ants' nest a brilliant ant-technology to produce more ants.

Here is a crucial point about the structure. Each oracle has to be related to a purpose. And a sequence of oracles will have to produce a coherent set of purposes just as evolution produces the coherent set of purposes to produce more grandchildren - the most straightforward measure of fitness. The problem with the "conscious network" is what produces the selection which gives purpose? The answer is the technological romanticists who desire to see a conscious network. That on its own guarantees nothing. Furthermore the structure which produces the oracles - hence the additional structure of the conscious network has to arise from the "ecology" - is the network as it is constituted at each stage of development. The object of Dr Frankenstein's vision is likely to be about as easily fulfilled with computer life as it was for Mary Shelley's original inventor.

Another source of scepticism can be derived from our complete ignorance of adaptive intelligence. We see ourselves as hugely intelligent in comparison to insects or lizards. This conceit fades when we're on a river with crocodiles. Alan Turing proposed a test of computer intelligence: a person asks questions of an agent whom he or she cannot see. The agent gives replies and after a sufficient number of questions and replies, the tester has to decide whether the agent was a computer or not. A discussion on whether this conforms to a sense of intelligence is a theme of the writings of John A. Searle (see for example *Minds, Brains and Science*, Harvard University

Press,1986). A criticism of the Turing test is that it is too easy and, indeed, might have been passed with the judge being Joseph Weizenbaum's secretary. In 1966 Weizenbaum produced a simple language processor and gave it the "domain" of a Rogerian Psychoanalyst. It would accept statements and paraphrase them as a question in the manner of a psychoanalyst intending to keep up the flow of meandering talk. Weizenbaum was shocked to find his secretary deep in serious conversation with the program. This program most likely would have passed the Turing test had the secretary been the judge.

I propose a different test.

Consider a lizard or crocodile. These large predators are hatched and come into the world with sufficient "hard wiring" to get themselves from where the egg is laid to water at an unknown location, and have to feed themselves and hide from predators which include others of their species. All this includes the ability to discriminate between large non-threatening objects such as trees and plants and the predators. They have to know when to hide and when to make a dash for it. They have to know what is food and what is not - day after day, month after month. They have to recognise, among their own species, who is likely to kill them, and with whom they should attempt to mate. No teachers anywhere. Initial brain size will be about the size of a crystal of coffee sugar.

The lizard test is to make a small robot, no bigger than an adult rat, with enough stored energy to get to water which might be 100m away. This should be done in an enclosed area with a large number of dogs, cats and birds which are trained to catch and retrieve the robot. (Strictly speaking the robot should carry a sign in dog, cat and bird lingo which says "eat me I'm tasty".) The environment this robot is let loose in is a dense forest or at least a well overgrown paddock.

Above all, no training of artificial neural networks is allowed.

When this is accomplished, Kurzweil and his ilk can announce the future promise of the age of lizard-intelligent machines.

Kurzweil himself estimates, on the basis of amounts of data processed, that IBM's chess playing program, Deep Blue, has the intelligence comparable to that of a lizard. This is a hard argument to make when intelligence is a property of behaviour which allows success in a variety of environments. And particularly hard when we have no idea how a lizard interprets its world.

For those who see this as too sceptical it not a theoretical consideration. Self managing networks are a goal of the computer industry. And indeed they will succeed in some measure, producing all sorts of techniques - such as self description - and experiments along the way. One of the projects, initiated by IBM, the *eLiza* project, is so called because of the alleged comparison between a lizard intelligence and Deep Blue. The goal leads to networks containing layers of algorithms that monitor the network and decide whether to spawn or initiate other algorithms. In this way it is plausible that the network has a "life of its own". But what we have is something that no longer is a device for human use. From the discussion in the previous paragraphs, this can only happen by human action. Just as we punish computer virus makers for making common programs and computers unworkable, we should be prepared to punish those who would seek to have networks become unpredictable in the attempt to "make them better". There is an analogy with genetic engineering. This new type of "engineering" must be treated with extreme caution. These experiments can lead to mixing of genetic segments which could possibly cause real problems with the horticulture of nations. Are the scientists prepared to accept responsibility for the bad effects of these technologies? Are the computer programmers who experiment with networks prepared to accept responsibility for the consequences of their actions? Certainly they can be sued. There are times when proposed improvements produce only marginal gains for what is little more than continued employment of specialists. These marginal gains are commonly experience by owners of personal computers who see only tiny benefits in upgrades of software often at the cost of having to buy more equipment. We require a little of the perspective of Luddites - a scepticism of the proposed improvements in a technology whose improvements have become detached from human purposes.

> "We will turn into robots. It's both inevitable and desirable." Robotics engineer Hans Moravec (as quoted in *The Australian*, 11/04/2000, IT Section p.20)

An interesting possibility is that the evolution of "conscious networks" can be stopped by sustained attacks of computer viruses. Indeed, developing network-stopping viruses could be humanity's last resort to control networks.

The age of spiritual machines foresees the merging of the human and the machine as people accept all sorts of implants and prostheses. This is plausible. As mentioned before, technological romanticists see interaction with machines as intrinsically good. A few generations of information-addiction will split humans into the "addicts" and the "rejectors". The addicts could easily avail themselves of devices to connect themselves permanently to various networks with tiny implants.

But what of complex networks? Can we expect levels of behaviour which we do not have words to describe? Is this descriptive lacuna already seen in ant colonies? Each ant is an autonomous communicating system within a thriving community of like-minded creatures. The ant colony as a whole exhibits characteristics different from and perhaps greater than the individuals. This would seem to be the case for ecosystems, cities and civilisations. The problem is what characteristics or properties do we look for in ant colonies, ecosystems and cities when we want to see them as alive or intelligent or for that matter thriving, decaying or dying? We actually do use these words in a metaphorical sense. Is there more to it? The puzzle recalls the Gaia controversy wherein the whole biosphere is taken as a living thing. Identifying characteristics of life or intelligence for disembodied, abstract, geographically dispersed entities is a problem. Even if some vast network were to be crudely "alive", how would we recognise it as so unless it communicated, or better still, conversed with us in a way which we recognised as a communication and an intelligent one at that? If the biosphere is intelligent and purposeful it would be nice to identify something of its language and actions. Would we recognise a biospheric chuckle? If the Internet is feeling clever or malevolent today will we ever know? Are such characteristics hopelessly anthropomorphic and do these creatures have "states of mind" unrecognisable, incomprehensible and indescribable like the Tao of Lao Tzu?

Some options on the answers are:

1. Intelligence is a characteristic of animals and has to be described as such. There is no evidence for anything remotely resembling intelligence in cities or civilisations which cannot be described in terms of agreements of humans to act in a certain way.

2. Even a huge network is, in computer science terms, a big "Turing Machine" or model of an abstract computer, and has all the limitations of such including the inability to deal with intractable problems - while humans actually can prove them intractable. This means mind is something other than a souped up Turing Machine.

3. We have a problem and we have to enter a new phase of humanity in which we recognise and ascribe new life-like

characteristics to networks and possibly ecosystems, cities and civilisations.

The future will answer.

Systemisation: The Irresistible Momentum

Underlying the changes that made the twentieth century a century of singular change, lay the rate of the increase in the abundance of goods. The ferocity of the two World Wars, the Cold War, the phenomena of popular mass broadcast entertainment and the various cults of personality large and small, the ownership of cars and the prevalence of international travel with its consequent growth of curiosity and tolerance, could not have happened without a society that produces vastly more than it needed. War is the ultimate waste. Goods are made to be demolished along with the lives of their targets. Mass entertainment, a cheap trickle at first but now the basis of huge industries and giant companies, requires mass consumers. Even de-colonisation owes something to the cheapness of radios and newspapers. Gandhi relied on British news networks to erode the British confidence in their moral superiority.

So much has happened, so much has changed our expectations, that we forget how much improvement - real progress - issued from such humble technologies as plumbing, flush toilets, electric lighting, heating, cooking and the telephone. Attention now is so riveted on computers and, more suspiciously, on biotechnology as the pinnacles of technological progress that it is easy to forget how recently they have arrived and how superfluous they are to the shape of the last hundred years. Plumbing of course is not a recent invention and electricity and the telephone belong to the nineteenth century. Their widespread use however, belongs to the twentieth. For the most part, the second half of the twentieth century saw the growth of guiltless consumerism. The generation which took part in the Second World War had to be coaxed into consumerism,

195

having been brought up in the depression of the 1930s. Their children accepted it readily and, in turn, their children enter adulthood with the expectation of credit cards and permanent prosperity. Each step in the abundance of goods becomes part of what is natural for those who have grown up with it. The rise in expectations becomes part of our culture in the sense of what is assumed, what you have no thought to challenge.

Culture is the set of expectations built on other expectations. In a rough form it is made up from building blocks. At the base is our understanding of what others expect of us and how we can gain status within the groups in which we live. Whom we might trust and whom we shouldn't trust. This forms the context of what obligations we ought to perform and what punishments are due to transgressions of obligations. This is the ethical and legal context of culture. These social aspects of culture arise from being a social animal. It is part of being what Aristotle saw as 'man the political animal'. From this springs expectations of what society has already provided for and what standard of care it provides. In the comfortable developed countries, we expect public hospitals, schools, roads, sewage and rubbish disposal. This is the arena of rising expectations. The final set of expectations - or hopes - are those of levels of tolerance and freedom from violence and arbitrary loss of freedom. From these hopes issue the model of the good life and how much leisure and pleasure it might contain. Here consumerism dominates. The luxuries of one generation become the rights of the next. And such rights are jealously held.

As de facto rights of consumption are established, no politician in a democratic country will speak against them. The maintenance of the general standard of living has become the primary aim of democratic politics. All else are side-shows. Politicians can embrace or flee from moral and technical conundrums but overall they are committed to growth. The fine tuning is all that is left. And fine tuning is all about international trade agreements; the flow of goods, the flow of money and expertise and labour, all of which diminish the autonomy of the nation state. The major styles of fine-tuning can be characterised in the first instance as putting individuals first, the uplifting and defence of individuals, the assumption that if you look after the people they will look after the "system". The opposing tendency is the feeling that the system provides the basis of security and wellbeing and all else flows from that. The second group is the bureaucrats of the world. These opposing styles of modern politics were discussed by Nigel Calder who made a whimsical analysis of the attitudes which characterised political groupings, regardless of what the parties might be called. Calder suggested the names "Mugs" (if the people are happy then there is nothing to be done) and "Zealots" (if the system is correct then there is nothing to be done). The following table, freely adapted from Calder's book, illustrates the idea.

Mugs	**Zealots**
Gandhi	Vladimir Lenin
Robert Kennedy	Henry Ford
Bill Clinton	Hitler
Jesus	Ronald Reagan
Mikhail Gorbachev	Winston Churchill
John Stuart Mill	Margaret Thatcher
Aristotle	Plato
Bertrand Russell	Pope John Paul II
Noam Chomsky	J. Edgar Hoover
Lao Tsu	Confucius
Edmund Burke	Robespierre
Einstein	Edward Teller
Abraham Lincoln	Louis XIV
Bill Joy	Hans Moravec
As a group, European Politicians	As a group, Asian Politicians
Jim Hacker from "Yes Minister"	Humphrey Appleby from "Yes Minister"

Mugs are liberals who get angry about who is not being nice to whom. Zealots get angry when their plans to enhance or maintain some system are derailed.

Zealots make the big rules and Mugs, when in power, do the fine adjustments which make society a kinder place. The Zealots have set the scene for the grand play of global supercompanies and the flood of goods. Mugs have liberalised society and produced our expectations of tolerance, schooling, health care and justice. The first relishes systemisation, the second desires the establishment of rules of equality and expectations of care and justice. These are "soft" systems, but they are systems all the same. Democratic principles are based on universal rules and universal rules play back into systemisation. They have to be applicable: circumstances have to be classified simply and unambiguously. Once classified as deserving, whether it is a tax rebate, a welfare payment, or a punishment for a crime, the amount which is due to a citizen has to be fair - standard for the circumstances. As in business, systems make things happen quickly and ease the conscience. And they are the answer to all problems. When things go wrong it is sensible to fix the immediate problem and then "put in place a system" to ensure the problem doesn't happen again or the beginnings of it are detected early. This applies to the control of burglars or bacteria, pests and perverts, currencies or criminals. Putting in place a system is the way we now solve or prevent problems.

When systemisation is a defence against disorder, truculence and culpability, society becomes an unhappier place. Intense competition, uncertainty and zealous systemisation will lead to a depletion of the spirit, a loss of generosity. Such is the irony of systemisation. It has led to an overabundance and all we can do is add to the abundance because systemisation has no social control.

Also ironic, amid this unprecedented abundance and prosperity, is an uneasy loss of control. If our world is a village - say of two thousand people - the adults will know each other. People are understandable and, even in the case of conflict, causes of misfortune can be ascribed. The villager knows "who has done what to whom". When the village grows to be a city, a person still might have a sense of belonging. "Local knowledge" still counts for something. Sacrifices can be endured because the benefits will accrue either to those making the sacrifice or their children. This feeling of belonging begins to wear thin at the national level and is very rare when life is enmeshed in global economics and business. There are still too few true citizens of the world. The bigger the system, and the more numerous its connections, the less it is understood. When poorly understood interactions take place thousands of kilometres away, and then affect us within days, we feel a loss of control. Global business does this all the time. Vaguely understood decisions of investors in foreign countries affect the prosperity of my neighbour and myself. An uneasy loss of control pervades the global economy because it is too big to see as something to which I contribute and which benefits me. The individual in a large system can be rescaled to be all but invisible. If you are invisible, you are very, very unlikely to make a difference. The tides of global economics wash over communities, leaving little sense of control or even knowledge of whom to blame. Awash with information, we decline in the control of our communities.

Contradictions

Wealth is the accumulation of goods. However and wherever they are produced, a rough measure of wealth is the quality and quantity of buildings, furniture, transport and leisure goods - all "things". This material abundance is produced by industries which use finance, marketing and planning specialists to institute change. These are in the information sector, a sector taken to include some 40 percent of the labour force. At the end of the twentieth century universities in the United States, employed more people than the manufacture of cars, aircraft, toys, textiles, plastics and chemicals. This is a dramatic example of our ability to pour out goods, the profits from which end up, among many other things, funding activities which are not immediately productive. This is independent of the source of the funds flowing into universities; whether they come via the government or not. This huge flow of money is a better indication of the change in employment patterns and expectations of expertise than slogans about the information society. This flow of money is happening at the same time as the growth in managed investment funds. It is indicative of the rise in productivity and the creation of surpluses in the last hundred years. Few of the graduates of the universities contribute to the acceleration of the production of goods. Within a few years of graduation, most of them will expect to be in management and administration, in services and in professions.

Seventy-five percent of the working population are now found in entertainment and tourism, education, health and "adjustment", business administration and marketing, government and local body administration, and the administration of utilities and infrastructure: water, energy, transport. Information technology has overwhelmingly been pushed along by administration. Its mission of creating and maintaining order is tantamount to a desire for systems. Mention systems and who should turn up but an Information Technology vendor. Administrators live in an abstract world in which each level of administrative decision making takes place in an atmosphere of increasing requirement for justification and performance. Every dumb or legally doubtful decision, however practical in the circumstances, attracts newspaper and TV reporters, angry parents and lawyers. Managing directors and administrators are fearful of trial by television where the disgrace of those in prestigious positions becomes popular entertainment. Popular entertainment, where heroes beat the system in show after show, becomes, with more than a touch of irony, the ultimate, if arbitrary, enforcer of systems.

Were we to have a stable state - which is indeed hypothetical - we might discover how to deploy a high percentage of the population in some socially acceptable work (which need not have anything to do with production). For long term prospects it is worth considering the extent to which the "adminosphere " is itself stable. Stability is probably the greatest of all threats to the adminosphere. Planning and decision making is predicated on change. This is true of a number of industries, especially the apparel business. Indeed, the highest level of planning might well be for social change which eliminates any threat of *stability*. This is reasonably easily accomplished; the vast amounts of money looking for promising investments feed changes to the skylines of cities and rearrange communities. Some of this disorients people and so provides for the growth of professions.

The professions in which the service amounts to people-being-with-people have the drawback of creating a society of disabled people. For reasons of employment, it might be desirable that any problem needs a counsellor, educator, investment advisor, health specialist, interior decorator, life-style planner..... As these groups argue that they are in the business of empowering people they are successful only insofar as they go out of business. The requirement for employment will most likely ignore this piece of logic and one of the changes will be the legitimisation of paralysis of choice and decision making outside a small, professionalised sphere of life.

The growth of the advisors and consultants is an inevitable outcome of complexity. No third-world village needs lawyers. Conflicts are resolved through commonsense or the authority of elders. As rules and choices proliferate, so do experts and specialists. This is an ordinary division of labour. The modern professions have arisen from the growth of rules, systems and products. Thus we derive our legislators and lawyers, accountants and water engineers, pilots and food technicians. Many thinkers believe that the idea of secure employment is something like the abundance of oil - an historical phase which will eventually end. The link between the current measure of economic well being, Gross Domestic Product, and jobs is becoming tenuous. "In the UK, between 1970 and 1995, GDP per head grew by nearly 60 percent but employment was 4 percent down. It was much the same in the rest of Europe. Even in America, which boasted of its new jobs, GDP grew by 50 percent in this period, while employment rose by only half that. In the weightless world of the new economy, it seems more than probable that countries can now get richer without more people working, or, more likely, with some working more, and more working less." Charles Handy predicts the growth of "Portfolio work , or what some have called 'women's pattern of life' will, in fact, become the norm for more and more people. They will increasingly combine bits of paid work with other forms of work, in the home caring for children or relatives, in the community for free, study work or even hobby work. It will be more meaningful in future to ask people what work they do than what job they have. More of us will lead 'actors' lives', alternating periods of project work with periods of 'rest and research', with temporary work or welfare to fill the financial gaps". Handy sees the workplace as involving "client contact". People will work as teachers have always done; work at the place of employment is with clients. It will be your responsibility to prepare for such interactions elsewhere. Work would be like a comfortable clubhouse as much for the clients as the employees. As has already been noted, this is happening with many people in highly regarded jobs in journalism, computers, engineering, marketing and in areas of businesses wherein the workers can be "virtual" with no fixed desk at work. They turn up at "the office" to pick up and deliver work or attend meetings.

Business can only expand by increasing the *busyness* in life. The writer of the long book doesn't consume much while writing. The person who intricately carves an ivory ball does not require many assets for that activity; they are not the consumers of entertainments. We are in a contradictory phase of civilisation. We have solved the production problems for most of the things we need. To retain employment we cannot capitalise on the cultivated leisure that this should bring. Leisure - which becomes an indicator of lifestyle - requires more assets, more goods. Entertainment and information are seemingly inexhaustible in their ability to be consumed. Inevitably they are THE growth industries.

The practical effect on large-scale planning is to draft information technology, in its broadest interpretation, into social changes which promote the demand for electronics. The telecommunications industry is adroit at this. The fear of missing a call from a customer or client has been turned into a preposterous duty to always be in touch. The pager, the cell phone - now items of fashion wear - are promoted as items of freedom and prestige, whereas when they first appeared many saw them as signs of slavery. Another example of changing culture to increase consumerism is to turn shopping from a social act to an information act. It is much more "efficient" for a "busy person" to consult elaborate hypertext catalogues over a network than actually make a quick choice in a shop and live with the choice. This importance of "choice" and the "right choice" increases the seriousness of the information part of shopping prior to the transaction. The one act of shopping has been expanded to sell more information services. Eventually it will expand to include on-line product consulting.

This information glut, this desire to be allowed to mull over all possible choices, this social duty to be informed or to be "in touch", is the consumer fashion of the information society.

It is a cliché that we work in an information society - that an increasing proportion of workers are information workers. And indeed an increasing proportion of workers works with computers, and computers have everything to do with information, therefore they are information workers. By the same token, anyone who typed letters, added up columns of figures or drew diagrams and planned, were information workers. For all their de rigueur fashion in business and government, the same activities go on except now they are mediated by the computer.

The core of information is to make distinctions. If you meet a lover in the park and it is agreed that a flower is to be worn in the lapel to signify some agreement, then the acceptance of the agreement is the information. The flower is merely a code.

Recall that information theory as developed by Shannon and others in the 1940s is code theory. How can letters be coded in the simplest way which allows them to be distinguished easily in spite of "static"? These are the concerns of information theory. But it is not the distinctions, the contexts of things, which are of concern.

The reason we live in a society rich in information is that there are lots of things to distinguish. There are more people doing more things with more goods. Computers, along with the media, telephones and faxes, bring to our attention items which are supposed to concern us. Electronics delivers information beyond anything that any person can deal with. Almost every item is trivial and transitory. To partake in this is to live trivially. It is to drive out reflection and thinking in favour of fashionable facts.

At the very end of the twentieth century, a movement called techno-realism started to publish books which warned of the effects of "infomania". While this movement is an important reaction to the unexamined information cult, it needs to be placed in the wider context. Information associated with the Internet or news media is part of consumption. No more or less significant than the glut of entertainment and less important than the corruption of politics to entertainment.

Inventions and patterns of consumption feed on each other. New inventions such as the mobile phone produce new markets which then have their own fashion and drive developments in mobile phone technology. There is no end in sight. Humans want to be Gods. They want to travel everywhere, effortlessly, and with maximum speed. So cars, aircraft, hang-gliding, deep-sea diving, low-orbit space travel. They want to communicate effortlessly with anyone they choose. So phones and the Internet. They want to understand each other's emotions - to beat the others and outguess them and have access to suspected concealed emotions of desired ones and opponents. Products yet to come. They want to explore, either passively or actively, and be challenged (a little bit). So travel documentaries and adventure holidays and computer games. They want to be fit, healthy and sexy forever. So pharmaceuticals, the health industry, hormone replacement therapy, Viagra and many products to come. They want to be well fed, comfortable and clean with safe havens. So restaurants, supermarkets, housing, furniture, security systems and insurance. Some, not many, are prepared to train hard to be highly skilled or accomplished. So training of all types. Each requirement is plied with products which move the consumer towards their desires. These products spawn lesser products and so on ad infinitum (well almost). The point is that this modest list of our immodest desires will fuel innovation for generations to come. Satiation is not in sight. Neither perhaps is wisdom.

Superabundance can only make sense if a culture of thrifty citizens gives way to heroic and flamboyant consumers living on credit. This shows itself in the rapidly increasing number of leisure related jobs. Shopping as the number one leisure related industry will be integrated as a total leisure package. Whatever the success of electronic commerce and Internet shopping, *real* shopping will be "experiential". The nature of management will be a reflection of a culture in which few people ever work with nature as in primary industries or working with metals, wood or ceramics. What is conceived as a problem and a solution will reflect the culture whose main preoccupation is having "fun."

In an age of superabundance the value of any tradable commodity will fluctuate and in most cases decline, as nothing tradable will continue to earn money forever. Not gold (substitutes can be found), not skills, they come and go. Everything is on the same ground as option trading. We now make these decisions all the time, we second guess trends, we manoeuvre to be trained in the most durable skills, investment thinking colours our choice of houses and children's schools. The basic mathematics of the age of superabundance is that of Black, Scholes and Merton and all the stochastic process mathematics. Until the Long Term Capital Management disaster these were hardly household names, well known only to option traders and barely known even among mathematicians. All of which is indicative of the specialisation and alienation which that brings. But these are names associated with the mathematics of fluctuations. They have provided option traders with the ability to skim off profits by gambling on the jagged and jittery graphs which track the fluctuations in currency and commodity values. But now commodities mean not only money and primary products but also components, technology, property and skills. Skills, particularly, will be subject to local shortages and gluts, increasing the movement of skilled workers around the world. But the ultimate currency is production expertise, followed by marketing expertise: expertise in materials (including agriculture, physics and chemistry) and production and supply management systems which determines the ultimate flow of wealth. All other wealth, however spectacular, as in the case of entertainers, sports performers, and some software and investment firms, is derivative.

Superproduction produces enormous concentrations of capital, and capital for new ventures is more abundant than ever. The fruits of abundance and good economic management now stream into pension and mutual funds which in turn go searching for the global "glamour stocks" so that fund managers can boast of high returns on investments. These fund managers, by their collective actions, create the stock market fashions and in so doing, transform the accumulations of the past into an appearance of future wealth. Whatever the current fashion, the consequence is that money is available to contemplate, plan and implement very large investments in technology: information superhighways, space stations, networks of satellites, and new levels of entertainment. Such technologies and entertainments need to strut the world stage in order to justify the investment in them. Global markets are a necessity. They can only be managed by elaborate, culture changing systems.

"They are not troubled by our problem with superfluity."

> Captain James Cook observing the Australian Aborigines indifference to his attempts at, and items of, trade.

Systems Thinking

Those who plan for tomorrow plant gardens
Those who plan for 10 years have children
Those who plan for 100 years plant trees

Chinese proverb

Concepts and expectations of government have been radically changed over the last 300 years. We have changed from accepting unconstrained rule issuing from a central hereditary group of privileged individuals to absolutely rejecting any such notion. Governments now make every claim to some type of democratic appearance. This change has not been linked to any particular technology. It has been a triumph of philosophy and political thinking; one of the few times in which the humanities can claim to have ameliorated the lives of people. But how much of this derives support from the rise of industry? The rise of factory work led to the need to educate workers in the technical operation of machines. The rise in standards of living, with its ready availability of books and newspaper to be read by a literate workforce, fuelled expectations. Finally the social mobility of the technically accomplished, mocked islands of privilege and snobbery and corroded class divisions. All these effects followed on systemisation. These changes, among the greatest achievements of humanity, involved many technologies, but have been driven by ideas of what a decent life should be for any human being. If it has been humanist ideas which have driven directions of social change, technology's role has been to open the gates. This is not a necessary connection, as the victims of too many perverted dictators can testify. But it is an overwhelming trend. The link between science and systems forces adherents of closed social, political or religious philosophies, such as communism or fundamentalism of various flavours, to compartmentalise their thinking. These compartments are imposed with fear on subject populations but crumble when fear is removed.

Any social change alters the way we exchange things. The accounting of the exchange of social obligations, news and goods, has become enmeshed in systems. Technologies always change the balance of exchanges. What was a craft, a chore or a burden, and hence accrued a level of gratitude, is devalued by mass production. Where technology breaks social conventions, as with contraceptives, the social effect can be dramatic.

Any exchange, in any system, takes place via a sequence of transactions according to some convention. The conventions form the grammar of social activities. These conventions interact with each other. They can follow on, reinforce, cancel or oppose, or otherwise interact in a meaningful "grammar" of actions. A transformation occurs when the relationship between classes of conventions changes. Then, certain sequences of interactions cease to be effective; rituals and appeasements fail, exchanges which usually reinforce bonds produce different reactions; what previously cancelled or excused, fails to do so. The sequences of actions are the fulfilment of general social expectations, what is rational in the minutiae of social intercourse, and this shifts, carrying with it those expectations. The spread of different forms of social expectations probably follows the same pattern of adoption of technological innovations. As with the adoption of technologies, changes in social expectations happen within a broad social setting and are constrained by a larger "logic" of ethics and custom. The link between the adoption of a technology and the simultaneous change in customs is demonstrated by the contraceptive pill. Before effective birth control, the ritual of marriage was most likely to be followed by the end of a woman's career. Now there is no such expectation. Society is changed at all levels. Expectations of life's cycles are changed and the possibility of making some plan for one's lifestyle is possible. Similarly, the speed, tactics and image of legal transactions have been changed by the fax. The slow movement of legal documents - allowing time for further consideration and scheming - is now no longer possible.

When social grammar, the sense of custom and norm, was taken seriously, high culture provided a break on mere merchants who usually ranked behind priests, government or court flunkies and warriors. Both in the East and the West merchants and traders have never been held in high esteem. The triumph of consumerism has swept all this away. The trader and manufacturer are king. And with this, so also their practical problem solving attitude and the embrace of systems - something not cultivated by the more "esteemed" occupations. Furthermore, science drew most of its foot soldiers from the merchant class. Instrumental thinking and system thinking are inseparable. If there is a problem and if there is a technological fix, let's do it. The inhibitions have gone. Birth control technology pushes aside cultural inhibitions; genetically engineered crops will find their market and solve their self defined problems. Custom now takes second place to the adoption of technology to solve problems. The systematic problem solving attitude is a great boon to humankind. It is much more effective in reducing misery than was custom and religion. It makes no claims to be universal, timeless or infallible. It has no exciting ring such as freedom or democracy and no media presence. But it is unromantically, supremely effective, and has transformed the lot of humanity more effectively than any religion or political movement. No one in East Asia doubts that, even though it is unnamed and formless as a modern Tao.

The rise of systems thinking has devalued a number of prestigious social skills just as literacy erodes careful listening and remembering. Once an educated person could recall thousands of lines of poetry. Such a person knew his or her culture: they were cultured or refined. These days we educate people to be operators in the social system. No one "in their right mind" does a history degree - it goes nowhere. A history graduate is neither use nor ornament except perhaps if they write for television. That history is the source of a broad perspective - the start of wisdom - does not matter. In a consumer society they can be a source of entertainment and that will do. But he or she does not fit into the system. They are not systems thinkers. Indeed all professions - from archivists and architects to weavers and zoologists - use the language of systems thinking, to some extent. Politicians and economists use it, and must do so. And so culture gives way to systems. The choice of a systems solution is dictated by global systems and the most widely recognised symbols are trademarks.

Culture is the source of much of our own mental inertia - but also the source of any sound judgement. We cannot invent future society without precedents and acknowledgements of the ultimate bedrock of culture: human nature. The cultural vacuity of the "digerati" vision of society; its aimless consumption and its incompleteness is illustrated from the passage by Steven Harnad of the University of Princeton, who described it elegantly as "rather as if Gutenberg and a legion of linotype operators, instead of Shakespeare and Newton, had determined what the printed page was to be used for." (*New Scientist* "Metropolitan" 25/3/95 p. 23). The word "digerati" was used (invented?) by Peter Thomas to describe the in-group enthusiasms for the digitised / wired / information society portrayed by Nicholas Negroponte in *Being Digital* (see *New Scientist* 25/3/95 p. 45). Michael Helm writing on "Virtual Reality" sees this vacuity as "infomania". That we become collectors of scraps, our attention span shortens - except, possibly, when glued to a computer screen - and we "..lose our feel for wisdom behind the knowledge." The common theme of these writers is that there is no clear reason for doing any of this. The announcements of the "digerati" seem to amount to more fun playing with computers and having access to more cable television and online shopping. Is this what it comes to?

Problems Solving versus Trading

Much has been accomplished by systems thinking. It has delivered abundance beyond blessing to a growing proportion of humankind. But at the turn of the millennium its contribution to solving social problems has been small. The future lies not in more goods, where there are already a surfeit, but in problem solving. This is an old theme. Lewis Mumford, writing in the 1960s, compared our civilisation to that of the Egyptian Pharaohs'. Ancient Egypt squandered huge resources on tombs. Monuments to death. Mumford saw this in the arms race. But he also saw that the consumer society, as obsessed with speed, automation and quantity without limit, had nowhere to go that made sense in truly human terms. He pleaded for a qualitative plenitude. This was not a going back to primitive strictures but transcending the obsession with glut and the vision of robot factories pumping out endless goods; a vision with no proper human goal in sight. In order to "forestall the ugly future that the prophets of megatechnics predict" it was necessary to take our models not from systems but from nature. Ecological and organic limits were required not so much to stop the sterilising of the Earth but to have lives that were meaningful.

The challenge is to focus the enormous intellectual energy which has produced the systems of abundance to produce the life which should issue from that abundance. Can we reset social agendas to solve problems differently, to allocate work sensibly, to increase what really makes for happiness: friendships, love, families, conviviality and plenitude - not surfeit? Above all there is a litany of growing woes. Human populations grow at the expense of wildlife; pollution has diffused throughout the biosphere even though the developed nations strive to contain and export their most obvious pollutants. Enormous health problems, largely water-borne and insect carried parasites and HIV, sap the underdeveloped world. There must be ways in which systems can help. However, this idealism must be tempered with the realisation that throwing money and techniques at problems does not always work. The world is seen differently by other people.

The abundance of the developed nations - the nations which have deployed systems most effectively - rests on a number of characteristics which we are only now beginning to see. Recall how mechanical systems force a scientific mode of thought. Explanations of failures have to be sought in the device itself, not in the intervention of fate or malevolence. Another potent effect of the increasing complexity of systems is faith in others - specifically experts in technology. Furthermore, this faith must be given without threat, beyond loss of face and income. Although expert metallurgists, artists, glassmakers and builders have existed in every rich society, most of these crafts have been easily explained to those who wish to employ them. Clocks and looms are less easily explained. Steam driven factories, chemical processes and electrical devices require trust in skilled people who, if bullied, can go to a competitor. Another aspect, already mentioned, for the successful deployment of systems, is a stable investment structure. This requires trust in institutions such as stock markets, banks and the legal system. All these aspects make up a set of social attitudes which are "normal" in the developed societies. Being normal we take them for granted. We do not see them ourselves; they are part of being an adult. But we do notice their absence in others.

There have been a number of books on the social attitudes which have contributed to the success first of all of the United States, Britain, with some of its colonial offshoots, the northern European nations and, more recently, of Japan, Singapore, Taiwan, Hong Kong and Malaysia. This is roughly a quarter of humanity. The question is why these people and not others? Why not the southern part of Europe and why not South America with its huge resources? There is no part of the old world where talent and organisation has not flourished at some time. There are no bars to talent in any nation so what is the reason? For much of the twentieth century the blame was laid on exploitation. The underdeveloped nations were that way because they were colonised and exploited. But then so were the working classes - now the middle classes of the rich nations. What allowed the most isolated modern nation, New Zealand, to develop as a stable modern nation while Uruguay, once a rich nation and with better geographical advantages, failed to maintain its status? The modern, post-colonial part of this exploitation explanation is international debt. Underdeveloped nations are said to be crippled by loans and foreign ownership. But loans were not given to produce nations of slaves but as investment in economies, and foreign ownership has been part of the training ground of nations. The loans were intended to be serviced by growing exports. The decisions to lend were based on the idea that people would use the money as the bankers would. That turned out to be the problem. Firstly corruption siphoned off the money to non-productive assets. For example from 1979 - 82 the siphoned flow is estimated as $US28 billion in the case of Mexico, $US21 billion from Venezuela, and $US12 billion from Argentina. All this means less money for investment while the whole debt is still to be paid off. It also imposes an effectively much higher interest rate on the uncorrupted part of the loan. There is no sense to invest when the recipients have no social infrastructure which can be trusted.

And this is a crucial part of modern attitudes.

From the 1970s, starting with a trickle of Latin American writers, the explanation of arrested development shifted from the exploitation explanation to an examination of culture. The Peruvian writer Mario Vargas Llosa summed it up for the Latin American case with, "The culture within which we live today in Latin America is neither liberal nor is it altogether democratic. We have democratic governments, but our institutions, our reflexes, our mentality are very far from being democratic. They remain populist and oligarchic, or absolutists, collectivist or dogmatic, flawed by social and racial prejudices, immensely intolerant with respect to political adversaries, and devoted to the worst monopoly of all, that of truth."

Francis Fukuyama in his book *Trust; The Social Virtues and the Creation of Prosperity*, puts the main differences in the wealth creating cultures to a wide "sphere of trust". Trust in people outside the family, trust in institutions such as equity markets. Fukuyama examines how in the United States, Germany, Japan and other developed countries, various institutions work together to build up manufacturing, marketing and banking conglomerates or communised economic institutions. The story is complex with many threads. In China, where Confucianism emphasised family loyalties and moral obligations fall off as one goes further from the immediate family, family owned organisations grow to the extent to which trusted related officials can control them. Of course all this is changing but the cultural component has delayed the creation of large organisations which can, as noted by A D. Chandler, gain the twin advantages of scale and scope.

Lawrence Harrison, author of *Culture Matters: How Values Shape Human Progress*, and whose conclusions are the same as those of Fukuyama, lists the main cultural obstacles towards progress as the resignation of the poor, an ideology of the virtue of poverty, low importance of education - especially for females - and a narrow sphere of trust - mostly confined to the family. His list of progressive attitudes is as follows. The concept of one's own destiny as being, to some extent, under one's own control, and that thrift, work, study and merit will be rewarded. One should add as well, a problem solving attitude; problems can be overcome and their overcoming a source of inspiration to others. A sense of fairness in society in which good work is rewarded and corruption punished. This goes together with rigorously applied social and legal codes and consequent trust in legally complying social institutions. That merit, and the constant achievement of desired results - rather than social connections - are central to advancement in society. A wide sphere of trust as described by Fukuyama. Authority is based on merit and allows dissent. The economic influence of what might be called centres of orthodoxy, usually religious institutions, is small; they might be big businesses themselves but their influence in business and thought is confined. A consequence, belatedly recognised, is a recognition that talent and merit are not confined to any class, race or sex.

These attitudes have grown up in Europe alongside growing systemisation. But nothing in Europe stands out as having been the germ of these attitudes. The Byzantine, or Eastern Roman Empire lasted for over 1,100 years as a Christian European entity and produced nothing other than buildings. The growth of systems thinking and scientific thinking has yet to be given a full account. But these are effective ideas. Aside from shouts of cultural hegemony and the tyranny of Western rationality all people want the results of these attitudes. They want to see all their children grow to take their place with some degree of status in society. And they want to live to see and play with their grandchildren. If these are rock bottom human desires, they are frustrated and statistically rare for a good proportion of humans who live in societies which do not have the attitudes listed by Harrison.

In the countries which have successfully adopted these attitudes we find the "middle classes". Progression into these middle classes is the best contraceptive. Middle classes tend to reduce the number of their children. They are also consumers - something the overproducing economies need - but the environment does not. They will readily reduce pollution if the means is not seriously inconvenient. The middle classes tend to have families with fathers of sufficient dignity to act as role models for sons. It would seem that among social animals, mature males, integrated in society with some status, are the best crime reduction mechanisms. Middle classing the world would solve a number of problems. This need not wash out all cultural differences but certainly some would go. The flaw in this plan is that middle classes, as a consuming group, rather than a group defined in terms of security, are depleting the Earth. The concept of an ecological footprint as the amount of land required to feed, water, house, treat the effluent and provide the energy for an individual, is useful as a measure of the environmental cost of a lifestyle. A middle class lifestyle can be taken as very roughly requiring four hectares. About 11 times what the average Indian requires. The concept which I have called middle classing is used to indicate certain types of responses which could be useful to solve some of the world's problems. Obviously consumption is not one of them. More radical ideas are needed for everyone. Another problem is that this implies a big change in culture and a change that would be resisted by those who stand to lose by it.

How can systems help? The following suggestions might be laughed at by the experts in the field but they are given as a stimulus to the imagination.

Frequent flier points and similar marketing devices testify to the ease with which we collect information about transactions. Well over two trillion dollars have been invested in entertainment companies and their associated infrastructure companies such as Microsoft and Cisco. These companies have been among the very top tier of capitalisation, all of which represents a huge surplus in the developed world. If every share transaction in such companies were taxed at one part per 1,000, this would yield a billion dollars which could go into reticulating water supplies in areas where it was practical. A drop but not nothing. Perhaps a one percent tax would not cause a riot, especially if seen as a long term investment in a stable and consuming world. Of course the trillion dollars a day which swims around the world could be taxed at one part per 10,000, which would involve some multiple taxing, and the tax could be directed for health work in desperate parts of the world.

Universities and polytechnics are now franchising courses and acting as corporations. Using the Internet and some mechanism of investment credits, staff could set up technical training centres in countries which are desperately short of even the first level of technicians. This would be the first step to capture the training markets where training is still in serious demand. The teachers would not need to be from the franchising body, so much as trained technicians willing to be involved and trained to deliver the material.

The overall contribution of systems now is in the collection of data in order to co-ordinate the movement of resources. The huge distribution problems which overwhelmed efforts to accelerate development can be solved. Given sufficient funding and effort, problems of development, habitat conservation, limiting pollution and growing and distributing expertise, can be solved. They are more challenging than working for Internet companies or whatever is fashionable. They will lead to new inventions and, with luck, new social perceptions. But they are also long term business opportunities. For the new ideas, the innovations in social arrangements and the use of systems will enrich our own culture of technique and problem solving. It will be a comic tragedy if the nations which have been the recipients of systems generated abundance have sunk so far into an insecure comfort that they do not even contemplate giving up the tiniest amount to secure improvements for which their grandchildren will thank them.

The twentieth century has been the century of manufacturing and trading. Social problems have been solved through systems but those systems have always been underpinned by trade. National status is measured by mastery of systems. Through them these nations have become autonomously productive and "modern". Nations as diverse as Russia, Japan, Malaysia and Iraq have adopted modern technology and the mastery of systems as the road to status and power. The resulting mode of operations has been that anything which facilitates trade is good, restrictions are bad. The whole world is a resource and entrepreneurialism is the highest social duty. This is in effect a "one solution - there is no alternative" approach to things. The Japanese government floated the idea that Japan should evolve from a trading nation to a High Technology Nation (as a way to solve resource insecurities primarily). Powerful trading groups buried the idea. But, indeed, we really need to evolve to problem solving nations. To leave the ideas of growth and look towards a dynamic plenitude (after Lewis Mumford). Our ability to produce goods in abundance should support the solution of social problems but not be seen as a solution itself. There are few social problems left which can be solved by increasing the supply of goods and hoping benefits will "trickle down". We need to use our understanding of systems - hopefully increasing all the time - to formulate better social systems for solving problems. Global trade will continue. But the challenge would be to solve problems at the primary level not via some diffusion of benefits. This evolution to a problem solving plenitude would be a fitting social goal for the potentially best educated and informed citizens who have ever lived.

Notes and References:

Author's caution
The Internet provided a number of references in this section. While this might be seen as a virtue there are two cautions. First is that there is no peer review of the accuracy of Internet information. Second is that the permanence of Internet pages is unknown. The author's expectation is that, as with other fashion and media dominated outlets, scholarly permanence and archiving are not high priorities.

Chapter 1

The history of technology, standardisation and systems comes from many sources. The *Encyclopaedia Britannica* (1984) (EB)is always a good source. But as information does not spring out from an open book, however large, the following sources have also been used.

Jean Gimpel's *The Mediaeval Machine*, Penguin 1976, (ISBN 0 14 00.4514 7) gives an important insight into the contribution of wind and water power to the development of industrialisation of the late Middle Ages.

The story of social systems and administration in China is worth a number of books. The best source to the author's knowledge is *The Shorter Science and Civilisation in China:1 and 2 An Abridgement by C. A. Ronan of Joseph Needham's original text.* Cambridge University Press 1978 ISBN 521 29286 7.

Hutchison's Dictionary of Scientific Biography has been useful for data on Babbage and EB is a source for Whitworth.

The data on James Watt, Babbage and Jacquard's loom came from Hutchison op. cit. and Donald Cardwell *The Fontana history of technology*, Fontana press 1994.

The Pinball Effect by James Burke, Little Brown and Company 1996, ISBN 0-316-11602-5 provided a number of additions to the story of James Watt.

The histories of Jacques de Vaucanson, Oliver Evans and Joseph Bramah are taken from *Mechanisation Takes Command* by Siegfried Giedion, Norton,1969, originally Oxford 1948. Some details of Bramah's, Maudslay's and Brunel's achievements are from *The Hutchinson Dictionary of Scientific Biography* op cit.

The Medieval Machine (op. cit.) p.15 describes "fulling" of cloth. The statistics on stitching come from *Encyclopaedia Britannica* (op. cit.) Industries, Manufacturing.

The rise of the American corporations in the late 19th and early 20th centuries is documented by Alfred Chandler in *The Visible Hand* and *Scale and Scope,* Harvard University Press.

The growth of industrial stocks on the New York Stock Exchange is recounted in John Steel Gordon, *The Great Game, A History of Wall Street*, Orion Business Books, London, 1999, p. 148.

Information on Frederick Taylor, Henry Ford and Alfred Sloan is from the *Encyclopaedia Britannica* (1984) and Anthony Sampson, C*ompany Man,* HarperCollins, London, 1995. The quote on Taylor is from p.43.

The production of the Ford Australia plant is from *The Australian* 18/11/99. The figure of 200,00 to 400,00 is from *Company Man* op. cit.

A tale of mechanisation including tractors and manufacturing wheat comes from *Mechanisation Takes Command* (op. cit.) Norton p.162.

The rise of United States business and finance in the First World War is from John Steel Gordon, *The Great Game, A History of Wall Street*, op. cit. p.203ff.

The importance of manufacturing technique and management skills to the ultimate outcome of the Second World War is told by Richard Overy in *Why the Allies Won*, Pimlico 1995, ISBN 0-7126-7453-5.

For a discussion on the tool investment see Lester Thurow, *Creating Wealth*, Nicholas Brealey Publishing, London 1999.

The statistics on risk swapping contacts is from *One World Ready or Not* by W. Greider, Penguin Press 1997.

Hewlett Packard spent $250 million developing the quiet printer and Gillette's "Sensor" cost $150 million to develop: New Scientist 21/10/89 p. 17.

The description of SABRE is from Meiklejohn 1988 *Airlines arm for on-screen war*, Business Computing and Communications February pp. 36-40. (Quoted from "*Business Computer Systems An Introduction*" D. M. Kroenke and K. A. Dolan, McGraw Hill 1989 p. 432.)

Much of the data on GATT can be found in *One World Ready or Not* by W. Greider, Penguin Press 1997.

The value of things in terms of wages was originally linked to the cost of food. See K. Polanyi, *The Great Transformation*, Beacon Press, 1957.

Oticon takes its product as *the perceived sound* as opposed to electronic amplifiers (New Zealand Herald 10/4/95).

The 3rd army alone evacuated by air 135,000 men in the last month of the war. Not a computer in sight. *The last offensive* of WWII by C. B. MacDonald 1973, ISBN 0 7924 5858 3.

Joseph Weizenbaum, Computer Power and Human Reason, MIT Press 1976 pp.28-29.

Langdon Winner, "Mythinformation in the High-tech Era" in T.Forester *Computers in the Human Context*, MIT 1989

For an idea of the philosophy of Postmodernism the reader could start with *Foucault* by J. G. Merquior, Fontana Modern Masters 1991, ISBN 0 00 686226 8.

Chapter 2

The vanishing of career prospects after 50 is examined in *Creating Wealth* op. cit. p.139.

The Sears book of rules is quoted in *Dangerous Company The Consulting Powerhouses and the Businesses that Save and Ruin* by Shea, J. O. and Madigan C. Nicholas, Brealey Publishing, London, 1997, p. 144.

The information on RAND Corporation and its activities is from Norman Moss, *Men Who Play God The Story of the Hydrogen Bomb*, Penguin Books, 1970.

Information about Magnetic Tape and Drum Storage was supplied to the author by Garry Tee of the University of Auckland in New Zealand.

The movement to increasing systemisation and automation is not without its critics. L. Hirschhoorn in an article "Robots Can't Run Factories" and J. Bessant and A. Chrisholm in "Human Factors in CIM" in the book edited by T. Forester *Computers in the Human Context*, MIT, 1989.

The productivity gap. See Chapter 8 "The Productivity Puzzle" in T. Forester ibid., and Scientific American November 1994 p. 83. Most recently T. K Landauer, *The Trouble with Computers, Useability and Productivity*, MIT, 1995.

"We see computers everywhere but in the productivity statistics," commented the Nobel Prize-winning economist Robert Solow (quoted from Paul A. David, "The Dynamo and the Computer: A historical perspective on the modern productivity paradox," American Economic Review, May 1990, p. 355).

See also William Bowen *The Puny Payoff from Office computers* in Chapter 8 "The Productivity Puzzle" in T. Forester op.cit.

The productivity paradox is worth a book on its own. A steady stream of articles and statistics exists on this subject. From *The Australian Financial Review* supplement "Boss" for 13/3/2000 the article "Who's in charge" by Emma Connors notes the following items:

- In 1996 some 73 percent of IT projects undertaken in corporate America were late, over budget or cancelled, according to the Massachusetts-based Standish Group. These project failures cost $US146 billion.

- The classic example of an Australian corporate folklore was the infamous CS-90 project at Westpac designed to produce a bank-wide banking platform. Four years after work began, CS-90 was abandoned in 1991 at the cost of $A150 million.

- John Thorp has been in the technology business since 1967. In his book *The Information Paradox*, Thorp points out that information technology budgets have been growing between 20 and 30 percent each year, "There is still no consistent statistical relationship between IT spending and various measures of value over time."

- Qantas' David Burden advises companies tempted to cut corners to think bridges: "You would never want to drive over a bridge that has been built by a bunch of amateurs without a design in a few weeks, but there is a tendency to believe that information technology systems, which are far more complicated than a bridge, can be thrown together in exactly that manner."

The Boston Consulting Group study of the dissatisfaction with software suppliers is quoted from the *Australian* 22/3/2000:

"The stagnant labour productivity…" is from P. Strassman in *The Information Payoff: The Transformation of Work in the Electronic Age* (1985), p. 163. The New Scientist 16th of August 1997 p. 13, cited a number of studies which showed that spreadsheets contained errors.

US National Bureau of Statistics figures on productivity are taken from Scientific American, May 2000 p. 20.

The productivity figures for service workers in comparison to manufacturing is from W. Bulkeley *Wall Street Journal* 1/3/93 and The Kanban comparison is from *Made in America*, MIT Press,

1989 (both quoted from Anthony Simpson, *Company Man; The Rise and Fall of Corporate Life*, Harper Collins, 1995).

The description of the Airline systems is from *Business Computer Systems An Introduction* D. M. Kroenke and K. A. Dolan, McGraw Hill, 1989 p. 432.

Shannon published his *The Mathematical Theory of Communication* in 1949. H. W. Kuhn defined information in terms of moves in a game in 1953. (Sources *Hutchison Dictionary of Scientific Biography*, 1994.) H. W. Kuhn was using a different definition of information in the mathematical theory of games and strategies. See *Classics In Games Theory*, ed. by H. W. Kuhn, Princeton Paperbacks, 1997.

And the promise of more amazing features are likely, according to W. W. Gibbs in an article *"Taking Computers to Task"*, Scientific American, July 1997, to make computers "…more fun and engaging to use. But will they earn their keep in the workplace?"

One need only look at even a child's book such as S. Beisty's and R. Platt's *Incredible Everything, How things are made,* Viking, 1997.

The sequence of stock market booms is from John Steel Gordon, *The Great Game, A History of Wall Street*, Orion Business Books, London, 1999, p. 58.

The 40 percent productivity for the 1920s is from *The Great Game* op. cit. p. 226 and the Depression years comes from, as does the U. S. corporate downsizing figures, Lester Thurow, *Creating Wealth,* Nicholas Brealey Publishing, London, 1999, p. xiv.

The banks, presumably with their worth based on accumulated surplus, "actual" wealth minus unpaid loans to customers, have had their contribution to the stock market index raised from 6 percent in 1987 to 22 percent in 1997. (Source The Adelaide Advertiser 18/10/97 p. 37.)

U. S. Mutual funds put a daily average of $US 7.5 billion into new economy stocks. (Australian Financial Review 25-26/3/2000 p 25.)

Chapter 3:

For the story of Chinese technology see Robert Temple, *The Genius of China*, Prion, 1998.

The story of Cheng Ho is from Encyclopaedia Britannica.

The story of many Hindu achievements in science such as the heliocentric solar system, ideas of atomism, rust-proof steel, and many others which predate the emergence of these ideas in the West is given in: http://www.hindubooks.org/sudheer_birodkar/india_contribution/astro.html

http://www.geocities.com/0080/Tokyo/Shrine/4287/education.html

http://hindunet.org/mathematics/geometrymaths.html

http://hindunet.org/sciences/

(Internet references supplied by Dr. Kamayasamy Surendran, Unitec Institute of Technology, Auckland, New Zealand.)

For Islamic achievements in science see Bloom, J. and Blair S., Islam, A Thousand Years of Faith and Power, TV Books, New York, 2000.

Karl Popper *The Logic of Scientific Discovery* (English Translation Hutchinson 1959).

"Sikorsky's *Ilya Mouremetz* (1914) stayed airborne for six and a half hours and carried six passengers". E. W. Montroll, *On the Dynamics and Evolution of Some Sociotechnical Systems*, Bull Am. Math. Soc. **16**,1,1987. According to Montroll the difficulty of developing a technology is related to the number of dimensionless constants associated with its processes.

The concept of falsification is due to Karl Popper - see Karl Popper, *The Logic of Scientific Discovery*, Hutchinson of London, 1959, or any of the numerous expositions of his work.

Details on Mendel taken from Encyclopaedia Britannica 1984.

The story of Thermodynamics and the Dynamo is taken from *The Fontana History of Technology* op. cit. and *The Dictionary of Scientific Biography* op. cit.

Tessa Morris-Suzuki in "The Technological Transformation of Japan", 1994, Cambridge University Press. The statistics on the rate of growth of scientists in Japan and of Riken are taken from chapter 5.

The story of Tomonaga Sin-itiro is told in *QED and the Men Who Made It,* by S. S. Schweber, Princeton Paperbacks, 1994.

Two references to Solar Storms are: http://www.abc.net.au/science/news/stories/s19873.html or http://www.partsonsale.com/solarstorms.html

Richard Pearse's flights in New Zealand in 1903 are documented in C. G. Rodliffe, *Wings Over Waitohi - the story of Richard Pearse,* 2nd edn. 1997, Avon Press, p.115. Also referenced in http://www.nzhistory.net.nz/Gallery/Pearse/Pearse.html. The affidavits of eyewitnesses include ample detail to show that Richard Pearse powered off the ground in 1902 and in 1903 several times before the Wright brothers' takeoff in Dec. 1903. None of these flights meets the definition of *controlled* powered flights, requiring lateral control to bring the aircraft to land near the takeoff, i.e. approx. 360° turn. Orville & Wilbur did not achieve this until 1905.

The story of Long Term Capital Management is taken from the script of a TV documentary, *The Trillion Dollar Bet.*

(http://www.pbs.org/wgbh/nova/transcripts/2704stockmarket .html)

See also http://www.numerix.com/ideas/offthewall/ for an article written at the time of the collapse of LTCM by Erik Aurell, Roberto Baviera, Maurizio Serva, Ola Hammarlid and Angelo Vulpiani, and "reports on their recent work which attempts to use large deviation theory to model rare fluctuations, while at the same time reducing to standard option pricing theory for small fluctuations."

The story of Jack Welch (GE) and the Taurus system are from Anthony Simpson, *Company Man; The Rise and Fall of Corporate Life*, Harper Collins, 1995, p. 218.

NASA's television spokesman's comments are reiterated in the Washington Outlook column of *Aviation Week and Space Technology,* 27/3/2000, p. 23.

Keith Allpress, of UNITEC Institute of Technology, pointed out that Le Chatelier's principle has numerous applications in organisations.

These ideas from ecology are associated with names such as the Odum brothers, MacArthur and Margelef - see M. Tanskey, "Structure, Stability and Efficiency in Ecosystems", in *Progress in Theoretical Biology,* ed. R. Rosen and F. M. Snell, Academic Press, 1976. Also Margelef, Ramon, "Diversity, Stability and Maturity in Natural Ecosystems", in van Dobben and Lowe McConnel eds., 1975, *Unifying Concepts in Ecology*, p. 54. Other ideas on the evolution of social-technology are in E. W. Montroll (op. cit.), particularly that the difficulty of developing a technology is related to the number of dimensionless constants associated with its processes.

Chapter 4:

Wilfred Hodges, Games and Forcing, 1984, provides a very technical reference to For All / There Exist games.

The application of game theory to auctioning of the electromagnetic spectrum is taken from '*John Nash and "A Beautiful Mind*"', John Milnor Notices of the American Mathematical Society, November 1998, p. 1,330. The Australian Government auction is reported in *The Australian,* 4/3/2000, p. 31.

The ideas behind the evolution of social structures (organisations) came from a series of papers published by the author: Dynamic Structures. An Algebraic Approach to Biological and Social Structures, *Bulletin of Mathematical Biology*, **43**, 5, pp. 579-591,1981. An Algebraic Approach to Taxonomy, Speciation and Niches. *Proceedings of the Second International Congress on Biomathematics*, 1984. *Aspects of Dynamic Structures,* Revue de Biomathematique, No. 86,1984.

Kurt Lewin's observation is quoted in Jay M. Shafritz and J. Steven Ott, *Classics of organisational theory,* 3rd Edition, Brooks (Cole Series in Public Administration, 1992).   For more information   on   Kurt   Lewin   1890-1947,   see   also http://www.geocities.com/~anchi/SAFT/whtlth.html

The nature of the predatory economy is documented in *One World Ready or Not.*

The data on Singapore's TradeNet is taken from *Corporate Information Systems Management: Text and Cases* by J. I. Cash, F. W. McFarlan, J. L. McKenney and L. M. Applegate, Harvard University Press, 3rd Edition 1992.

The economic role of the *Cheobal* and *Ziabatsu* is detailed in Morris Suzuki op. cit. p. 164

Edmund Burke's *Reflections on the French Revolution* is a defence of custom and tradition as social guides.

Chapter 5

More information on industrial disasters can be found from TED, the Trade and Environmental Database available from: http://www.american.edu/projects/mandala/TED/class/all.html

Likewise more information on aircraft disasters can be found from the reports associated with the listing of aircraft accidents on http://dnausers.d-n-a.net/dnetGOjg/Disasters.html. The Sioux city crash report, Report No. NTSB-AAR-90-06 was found through this web-address. For detailed accounts, both the *Air Disaster* volumes by Macarthur Job, Aerospace Publications, and *Black Box* by Nicholas Faith Boxtree, 1998, tell the story of little things mounting to catastrophes.

The outline of the Mt. Erebus crash is taken from *Air Disaster* Volume 2.

The statistics for accidents per miles flown is from *Losing Altitude,* by Steve Creedy, *The Australian* 4/3/2000.

The development of propulsion control of aircraft PCA systems is reported in Scientific American, June 2000, p.16.

The loss of skills at NASA is reported in *NASA's not shining moments*, by J. Olberg, Scientific American, February 2000.

The report on the loss of skills in the Australian public sector is from *The cost of contracting in the dark,* by A. Yates, *Australian Financial Review*, 3/6/2000.

By some accounts, 40 percent of large projects fail in some way or are cancelled. These kinds of statistics are found in any book on project management. This one came from *MIS Australia,* November 1998, p. 48.

Chapter 6

Almost any book on Western philosophy will discuss these ideas of Hume, Husserl and Kant. See e.g. *The Penguin History of Western Philosophy,* by D. W. Hamlyn, 1987.

A simulation of thousands of coloured point objects which interacted and changed colour according to simple rules produced no clear pattern for thousands of cycles. Then spirals of colour started to form and rotate. What type of abstraction will deliver an explanation? See D. Griffeath, *Cyclic Random Competition. A Case History in Experimental Mathematics,* Notices of the American Mathematical Society, Vol. 35, No. 10, Dec. 1988, p.1,472.

Ray Kurzweil's *The Age of Spiritual Machines, When Computers Exceed Human Intelligence,* Viking Press, 1999, is the source book of visions of the final embrace of humanity by machines. A similar line of thought is in Hans Moravec, *Robot: Mere Machine to Transcend the Mind,* Oxford University Press,

1998. The classic of the criticisms of machine intelligence comes from one of its pioneers, Joseph Weizenbaum, cited above, and John A Searle's *Minds, Brains and Science*, Harvard University Press,1986.

The content and opposing styles of modern politics were discussed by Nigel Calder, in his book *Technopolis,* Panther, 1970.

Charles Handy, *The New Alchemists*, with photographs by Elizabeth Handy, Hutchinson, 1999. The statistics on GDP and employment are from the review in *The Irish Times*, Monday, November 22, 2000.

"Higher education...is larger than automobile, aircraft, textile, construction machinery, mining machinery, toy and sporting goods, household video and audio, and refrigeration and heating equipment industries combined in terms of direct employment." Congressman G. E. Brown Jr., *What is the Future for the Physical and Mathematical Sciences,* Notices of the American Mathematical Society, Vol. 42, No. 7, July 1995, p. 765.

A transformation is like a change in the rules of the grammar of social activities. Some sequences cease to have meaning. This is meant to be analogous to Noam Chomsky's deep grammar - see e.g. Lyons, J., *Chomsky* Fontana / Collins 1971. The spread of different forms of social expectations probably follows the same pattern of adoption of technological innovations (see E. W. Montroll, op. cit.). The classic study of the change of social patterns is in Cavalli-Sforza, L. L., and Feldman M. W., *Cultural Transmission and Evolution, A Quantitative Approach*, Princeton University Press, 1981.

At the very end of the 20th century, a number of books were published which were critical of the glut of information: *"Faster"* by James Gleick, *"Coercion"* by Douglas Rushkoff, and *"The End of Patience"* by David Shenk — which purport to expose the dark underbelly of technology. Less doom-and-gloom neo-Luddites than "enlightened skeptics" (as Shenk's jacket-flap puts it), Gleick, Rushkoff and Shenk in *Is technology unplugging our minds?* examine how the information overload is affecting us, changing our lives and rewiring our brains.

Michael Heim, The Metaphysics of Virtual Reality, Oxford University Press, 1993.

Lewis Mumford, The Pentagon of Power, The Myth of the Machine, Harcourt, Brace and Jovanovich 1970, p.402.

Gerald Segal, The World Affairs Companion, Simon and Schuster, 1996, gives the figures for corruption in Latin America.

Francis Fukuyama, Trust The Social Virtues and the Creation of Prosperity, Penguin Books, 1995.

Lawrence E. Harrison, Culture Matters, *Australian Financial Review*, 3/11/2000 and Culture Matters: How Values Shape Human Progress, Basic Books, 2000.

The figure of the ecological footprint for middle classes is taen as a guess from that published for the average Australian, 4.4 hectares. *The production time bomb*, Peter Fisher, *Australian Financial Review*, Review section, 25/2/2000.

INDEX

About the Author

Andrew Macfarlane has worked in the computer industry since the 1974. He has been a systems analyst and programmer, an IT manager in manufacturing and service organisations, a consultant and a lecturer. In addition to his involvement with business systems he has taught courses on aspects of technology and society and on Western philosophy. He has also written technical papers on Quantum Mechanics and Mathematical Biology. He lives in Gold Coast City in Australia and works as a business analyst and systems tester.

www.ingramcontent.com/pod-product-compliance
Lightning Source LLC
Chambersburg PA
CBHW020741180526
45163CB00001B/310